ENVIRONMENTAL SCIENCE, ENGINEERING AND TECHNOLOGY SERIES

ECO-CITY AND GREEN COMMUNITY

THE EVOLUTION OF PLANNING THEORY AND PRACTICE

ENVIRONMENTAL SCIENCE, ENGINEERING AND TECHNOLOGY SERIES

Nitrous Oxide Emissions Research Progress
Adam I. Sheldon and Edward P. Barnhart (Editors)
2009. ISBN: 978-1-60692-267-5

Fundamentals and Applications of Biosorption Isotherms, Kinetics and Thermodynamics
Yu Liu and Jianlong Wang (Editors)
2009. ISBN: 978-1-60741-169-7

Environmental Effects of Off-Highway Vehicles
Douglas S. Ouren, Christopher Haas, Cynthia P. Melcher, Susan C. Stewart, Phadrea D. Ponds, Natalie R. Sexton, Lucy Burris, Tammy Fancher and Zachary H. Bowen
2009. ISBN: 978-1-60692-936-0

Agricultural Runoff, Coastal Engineering and Flooding
Christopher A. Hudspeth and Timothy E. Reeve (Editors)
2009. ISBN: 978-1-60741-097-3

Agricultural Runoff, Coastal Engineering and Flooding
Christopher A. Hudspeth and Timothy E. Reeve (Editors)
2009. ISBN: 978-1-60876-608-6 (Online Book)

Conservation of Natural Resources
Nikolas J. Kudrow (Editor)
2009. ISBN: 978-1-60741-178-9

Conservation of Natural Resources
Nikolas J. Kudrow (Editor)
2009. ISBN: 978-1-60876-642-6 (Online Book)

Directory of Conservation Funding Sources for Developing Countries: Conservation Biology, Education and Training, Fellowships and Scholarships
Alfred O. Owino and Joseph O. Oyugi
2009. ISBN: 978-1-60741-367-7

Forest Canopies: Forest Production, Ecosystem Health and Climate Conditions
Jason D. Creighton and Paul J. Roney (Editors)
2009. ISBN: 978-1-60741-457-5

Soil Fertility
Derek P. Lucero and Joseph E. Boggs (Editors)
2009. ISBN: 978-1-60741-466-7

Handbook of Environmental Policy
Johannes Meijer and Arjan der Berg (Editors)
2009. ISBN: 978-1-60741-635-7

The Amazon Gold Rush and Environmental Mercury Contamination
Daniel Marcos Bonotto
and Ene Glória da Silveira
2009. ISBN: 978-1-60741-609-8

Buildings and the Environment
Jonas Nemecek and Patrik Schulz (Editors)
2009. ISBN: 978-1-60876-128-9

Tree Growth: Influences, Layers and Types
Wesley P. Karam (Editor)
2009. ISBN: 978-1-60741-784-2

Syngas: Production Methods, Post Treatment and Economics
Adorjan Kurucz and Izsak Bencik (Editors)
2009. ISBN: 978-1-60741-841-2

Syngas: Production Methods, Post Treatment and Economics
Adorjan Kurucz and Izsak Bencik (Editors)
2009. ISBN: 978-1-61668-214-9 (Online Book)

Process Engineering in Plant-Based Products
Hongzhang Chen
2009. ISBN: 978-1-60741-962-4

Potential of Activated Sludge Utilization
Xiaoyi Yang
2009. ISBN: 978-1-60876-019-0

Recent Progress on Earthquake Geology
Pierpaolo Guarnieri (Editor)
2009. ISBN: 978-1-60876-147-0

Clean Fuels in the Marine Sector
Environmental Protection Agency
2010. ISBN: 978-1-60741-275-5

Clean Fuels in the Marine Sector
Environmental Protection Agency
2010. ISBN: 978-1-61668-431-0 (Online Book)

Freshwater Ecosystems and Aquaculture Research
Felice De Carlo and Alessio Bassano (Editors)
2010. ISBN: 978-1-60741-707-1

Species Diversity and Extinction
Geraldine H. Tepper (Editor)
2010. ISBN: 978-1-61668-343-6

Species Diversity and Extinction
Geraldine H. Tepper (Editor)
2010. ISBN: 978-1-61668-406-8 (Online Book)

Harmful Algal Blooms - Impact and Response
Vladimir Buteyko
2010. ISBN: 978-1-60741-665-4

Estuaries: Types, Movement Patterns and Climatical Impacts
Julian R. Crane and Ashton E. Solomon (Editors)
2010. ISBN: 978-1-60876-859-2

Built Environment: Design, Management and Applications
Paul S. Geller (Editor)
2010. ISBN: 978-1-60876-915-5

Wildfires and Wildfire Management
Kian V. Medina (Editor)
2010. ISBN: 978-1-60876-009-1

HYDRO GIS: Theory and Lessons from the Vietnamese Delta
Shigeko Haruyama and Le Thie Viet Hoa
2010. ISBN: 978-1-60876-156-2

**Grassland Biodiversity: Habitat Types, Ecological Processes
and Environmental Impacts**
Johan Runas and Theodor Dahlgren (Editors)
2010. ISBN: 978-1-60876-542-3

**Check Dams, Morphological Adjustments and Erosion Control
in Torrential Streams**
Carmelo Consesa Garcia and Mario Aristide Lenzi (Editors)
2010. ISBN: 978-1-60876-146-3

Fluid Waste Disposal
Kay W. Canton (Editor)
2010. ISBN: 978-1-60741-915-0

Pollen: Structure, Types and Effects
Benjamin J. Kaiser (Editors)
2010. ISBN: 978-1-61668-669-7

**Psychological Approaches to Sustainability:
Current Trends in Theory, Research and Applications**
*Victor Corral-Verdugo, Cirilo H. Garcia-Cadena
and Martha Frias-Armenta (Editors)*
2010. ISBN: 978-1-60876-356-6

Environmental Modeling with GPS
Lubos Matejicek (Editor)
2010. ISBN: 978-1-60876-363-4

Wood: Types, Properties, and Uses
Lorenzo F. Botannini (Editor)
2010. ISBN: 978-1-61668-837-0

Mechanisms of Cadmium Toxicity to Various Trophic Saltwater Organisms
Zaosheng Wang, Changzhou Yan, Hainan Kong and Deyi Wu
2010. ISBN: 978-1-60876-646-8

Eco-City and Green Community: The Evolution of Planning Theory and Practice
Zhenghong Tang (Editor)
2010. ISBN: 978-1-60876-811-0

Anthropology of Mining in Papua New Guinea Greenfields
Benedict Young Imbun
2010. ISBN: 978-1-61668-485-3

Paleoecological Significance of Diatoms in Argentinean Estuaries
Gabriela S. Hassan
2010. ISBN: 978-1-60876-953-7

ENVIRONMENTAL SCIENCE, ENGINEERING AND TECHNOLOGY SERIES

ECO-CITY AND GREEN COMMUNITY

THE EVOLUTION OF PLANNING THEORY AND PRACTICE

ZHENGHONG TANG

EDITOR

Nova Science Publishers, Inc.

New York

NOTICE TO THE READER

The Publisher has taken reasonable care in the preparation of this book, but makes no expressed or implied warranty of any kind and assumes no responsibility for any errors or omissions. No liability is assumed for incidental or consequential damages in connection with or arising out of information contained in this book. The Publisher shall not be liable for any special, consequential, or exemplary damages resulting, in whole or in part, from the readers' use of, or reliance upon, this material.

Independent verification should be sought for any data, advice or recommendations contained in this book. In addition, no responsibility is assumed by the publisher for any injury and/or damage to persons or property arising from any methods, products, instructions, ideas or otherwise contained in this publication.

This publication is designed to provide accurate and authoritative information with regard to the subject matter covered herein. It is sold with the clear understanding that the Publisher is not engaged in rendering legal or any other professional services. If legal or any other expert assistance is required, the services of a competent person should be sought. FROM A DECLARATION OF PARTICIPANTS JOINTLY ADOPTED BY A COMMITTEE OF THE AMERICAN BAR ASSOCIATION AND A COMMITTEE OF PUBLISHERS.

LIBRARY OF CONGRESS CATALOGING-IN-PUBLICATION DATA
Eco-city and green community : the evolution of planning theory and practice / editor, Zhenghong Tang.
    p. cm.
Includes index.
ISBN 978-1-60876-811-0 (hardcover)
1. City planning--Environmental aspects--Case studies. 2. Urban ecology (Sociology)--Case studies. I. Tang, Zhenghong.
HT166.E326 2009
307.1'216--dc22
                                    2009048933

Published by Nova Science Publishers, Inc. ✦ New York

CONTENTS

PREFACE

In the 21st century, urban planners are faced with many new challenges which include climate change, ecosystem degradation, loss of biodiversity, community vulnerability, and social inequalities. The concepts of "green community" and "eco-city" have emerged as important integrated approaches for achieving environmental and economic sustainability at the local level. However, the idea of and need for eco-city and green community development and the values connected to this topic have not been sufficiently studied as far as the wider public is involved. Urban decision makers, planners, stakeholders, and the general public are all confronted with a lack of understanding of the concepts and implementation of this topic. Therefore, it has become urgent and necessary to make an innovative study of the theory and practice of eco-city and green community.

The principles and techniques of eco-city and green community build a feasible bridge leading to a healthy, diverse, balanced natural and human environment that is necessary for long-term sustainable development at the local and national levels. Communities featured as "green community" and "eco-city" can provide a range of solutions to current urban planning problems of urban sprawl, environmental pollution, land contamination, exploitation of natural resources, unhealthy built environment, and unequal social development. The core of this new approach is to build a sustainable inter-relationship between physical environment and human settlements. There is an increasing consensus among researchers and planners that the new form of eco-city and green community has significant positive effects on urban growth, environmental protection, and sustainable development. Demand for guidelines for eco-city and green community is growing and urgent as planners review the weaknesses of traditional urban forms.

This book "*Eco-City and Green Community: The Evolution of Planning Theory and Practice*" provides a comprehensive overview of the evolving theories and practices of "green community" and "eco-city." It explores the role of urban planning and offers an historical review of important problems which deal with the theories and applications of eco-city and green community. As a rapidly evolving concept, researchers and planners are paying more attention to the theories, principles, and practices of eco-city and green community. This book provides core themes, principles, and cases to discuss this topic as well as insight to future prospects for eco-city and green community development.

The three main sections in this book advance our understanding for the theory and practice of eco-city and green community. The first section explains the theoretical foundation of eco-city and green community development and serves as a basis for new planning forms to maximize quality of life and living in harmony with the environment. Second, the approach section offers a wide range of strategies for building green community and advancing the eco-city agenda. It further links the utopian idea of eco-city and green community to effective planning policies. The strategies provide an overview of traditional urban planning areas in which the ecological planning approach stimulates rethinking or restructuring of existing planning decision processes. Third, the practice section illustrates an international view for existing models of green community and eco-city. Real cases examine the vision and concepts of eco-city and green community and provide inspiration and guidelines for the development of sustainable, eco-friendly, green societies. These cases show how the standards of eco-city and green community can be reached within the existing planning context that has traditionally focused on short-term values. The book also points out potential obstacles which can delay or hinder the implementation of the envisioned eco-city and green community.

This book will significantly challenge established planning models and existing design standards and provide new ecological perspectives and sustainable inspiration for long-term practical solutions. The findings of the book expand existing major urban planning theories by taking broad theoretical principles and examining them in a practical context to illustrate their effectiveness.

Wayne Drummond
Dean, Professor, College of Architecture
University of Nebraska-Lincoln

ACKNOWLEDGEMENT

I am sincerely grateful for the generous support I received from Professors Wayne Drummond, Gordon Scholz, and Katherine Ankerson which enabled me to complete this book. Special thanks also to the team workers and co-authors who provided special collaboration.

This research is supported by a Hyde Professorship Fund and Layman Award from the University of Nebraska-Lincoln. The findings and opinions reported in this book are not necessarily endorsed by the funding organization.

Zhenghong Tang
Lincoln, Nebraska

SECTION I: INTRODUCTION

In: Eco-City and Green Community
Editor: Zhenghong Tang

ISBN: 978-1-60876-811-0
© 2010 Nova Science Publishers, Inc.

Chapter 1

INTRODUCTION

Zhenghong Tang and Ting Wei

Community and Regional Planning Program
University of Nebraska – Lincoln, NE, USA

PROBLEM STATEMENT

The world has been significantly urbanized since the 1900s and world population has grown rapidly in the last century and will continue this trend in the future. Although population growth rate of developed countries is increasing relatively slowly, developing countries face multiple pressures from urban population growth, economic explosion, and land consumption at a terrifying rate. Rapid population growth, global economic expansion, and massive urbanization have caused an exponential growth in the volume of resources consumed and pollution created around the world. Yet, perversely, a significant portion of the growing urban population lives in low-quality communities with no running water, no electricity, no sanitation, no clean air, no health care, no shelter, no transit, no public space, and there are millions who are still living in life-threatening urban environments or unhealthy urban communities. Environmental decline continues to accelerate in many cities in the long-term. Many city dwellers are threatened by overwhelming pollution and unequal community. Since the 1970s, car-dependent development patterns have caused rapid urban sprawl whose effects on local communities include the unavoidably negative conversion of productive agricultural and ecologically fragile lands into built environments such as residential housing and auto-dominated commercial strips and highways. Urban sprawl significantly converts natural lands to built environment, and creates huge pressures on environmental protection, urban infrastructures, urban services, and social equality. Not only does urban sprawl lead to direct or indirect environmental degradation, but, more importantly, it can cause social problems such as racial segmentation and environmental injustice.

In the 21th century, urban planning is facing many challenges and opportunities. In general, rapid urban development and population growth have contributed greatly to serious environmental burdens. Our cities and communities are confronted with many unavoidable environmental and social problems including air pollution, environmental degradation, transportation congestion, health issues, global warming, environmental injustice, and social inequity. Our communities and cities, as mankind's habitat, caused major destruction of the natural environment when creating the living environment. Urban development threatens the ecosystem and human survival. Today's cities consume most of the world's energy and cause significant waste and pollution. Urbanized areas are the centers of production and consumption of most non-agricultural goods and materials. Cities are suffering endless pollution, waste, overcrowding, and inner-city decay (Girardet, 2004). While urban problems are defined broadly, it can be illustrated that urban environmental problems threaten present and future populations, by human-induced damage to the natural and built environments. The sources of urban environmental problems include: 1) localized community environmental problems generated by the community itself such as noise, indoor air pollution and household waste water; 2) urban-regional environmental problems that may mainly affect the city, such as river water quality; and 3) regional-strategic environmental burdens that have cross-boundary, long-term impacts, such as biodiversity, ecosystem, and climate change. As many environmental problems require community participation, an increasing number of voices are calling for more locally-driven environmental approaches. It is important to link abstract environmental problems with responsible local initiatives in a coherent fashion. In short, it would seem that helping cities to address the goals of eco-city and green community will contribute to sustainable development.

All these problems provide opportunities for planners to create and maintain healthy environments and natural systems while conserving energy, water, soils, and important natural resources and reducing waste and cleaning up pollution and brownfield sites. Some planners have made significant progress in infill development, transit-oriented development, and compact development. The principles of mixed-use commercial and residential development have been accepted in many cities around the world. Environmental justice has become an important concern in the siting of controversial land use projects. Urban planning has paid special attention to the protection of sensitive natural areas and the creation of open green spaces. In short, urban planning has become an important element in addressing environmental challenges.

Planning emerges out of a series of crises, such as health crises (e.g. diseases), social crises (e.g. riots), and other crises (e.g. hazards). In the 19th century urban public health became a focus of concern in the planning field. In 1876, Benjamin Ward Richardson published an important book "*Hygeia, City of Health.*" This book proposed that planning should address air pollution control, water quality purification, sewage treatment, public laundries, public health inspectors, and other important health issues. The concerns proposed in this book stimulated the Parks Movement in the 19th century which grew out of landscape architecture, garden design, and community design. It emphasized shifting planning from private to public settings. Frederick Law Olmsted, also a key person in the parks movement, designed numerous park projects including Prospect Park, Chicago's Riverside subdivision and World's Fair design, Buffalo's park system, the park near Niagara Falls, and the Boston park system. His works integrated "naturalistic" design or "organic" design and addressed social-psychological impacts. The milestone of the parks movement is Central Park created in 1857 in New York City. The parks movement is the original starting point of eco-city and green community. The goal of the parks movement was to separate transportation modes, support active and passive uses of urban lands, collect urban water, and promote moral pastimes. Many of those ideas are still important topics in current planning for eco-city and green community development.

The Garden City movement was an important element in the development of eco-city and green community. Many researchers believe that this movement was ad hoc, but the Garden City movement was indeed planned. The concepts and principles in this movement greatly influence the development of modern city planning which rapidly spread from Europe to the United States and other countries. Ebenezer Howard was the key person in the Garden City Movement which proposed three concepts: 1) town level with higher wages, many jobs and amusement; 2) country level with natural beauty, low rent, and good air quality; 3) town-country level with a combination of the above. He emphasized using greenbelts to separate the central city which would be an ideal, self-contained city within the core area and the surrounding greenbelt. This idea is widely used in many countries to discourage metropolitan sprawl and industrial centralization. In England, his Garden City ideas were implemented in two towns, Letchworth and Welwyn in the early 1990s. Currently hundreds of cities throughout the world claim to be garden cities. However, many of Howard's original Garden City ideas have failed to be implemented. Many residential homes have been developed in suburbs causing urban sprawl with local residents commuting to work from suburbs to downtown because local industries and businesses do not provide adequate job opportunities.

The sustainable development concept emerged in the 1980s. No event did more to push sustainable development into the mainstream of worldwide policy debates than the 1987 release of the report of the World Commission on Environment and Development, commonly known as the Brundtland Commission. The Bruntland report for the U.N. defines sustainable development as *"development that meets the needs of the present without compromising the ability of future generations to meet their own needs"* (The World Commission, 1987). The report succeeded remarkably well at calling global attention to the need for sustainable development and developing a common formulation of this concept. The second conference, held in Rio de Janeiro, in 1992, produced a lengthy declaration known as Agenda 21 that laid out sustainable development principles in many different areas. Moreover, it raised a point of "Local Agenda 21." This mandate for "Local Agenda 21" planning has stimulated a large number of local planning initiatives, especially in the U.S., Europe, and a number of developing nations. In the planning field, Philip Berke and Maria Conroy (2000) defined sustainable development as *"... a dynamic process in which communities anticipate and accommodate the needs of current and future generations in ways that reproduce and balance social, economic, and ecological systems, and link local actions to global concerns."* Sustainable development emphasizes harmony with nature, livable built environment, and local-based economy. The dream of eco-city and green community has been included as a part of the sustainable development campaign.

The concept of eco-city and green community is different from traditional planning models since it incorporates important ecological principles in its design and functions. The idea and techniques of eco-city and green community can help communities understand how to become more sustainable by changing behaviors that contribute to resource utilization and conservation. Eco-city and green community can maximize the quality of life, minimize total lifecycle costs, and reduce transportation demand. The implementation of eco-city and green community covers urban structure, transportation systems, energy and material flows, and other socioeconomic sectors. Urban planning for eco-city or green community provides a general framework for understanding the urban function of the natural and built environment. Urban planners for eco-city or green community need to consider that the physical structure of the community is an interconnected system which includes land demands, land use types, urban growth boundaries, open spaces, and density. Transportation systems are a major factor in the development of eco-city and green community and cover physical and virtual movement of people, goods and data. Energy system has become an important concern in this campaign which addresses the movement or flow of energy and materials in the urban system. Last, the socioeconomic context includes human and economic activities that relate to urban growth.

Perhaps the eco-city and green community effort is not the only way to solve urban environmental problems; but it is probably one of the most likely ways to lead to a solution for a future sustainable society. The concept of eco-city and green community has been proposed as the backbone for sustainable development in the United Nations report of "Our Common Future" (1987). The idea of eco-city and green community addresses the core concept of sustainability and redefines wealth with natural capital such as clear air, fresh water, clean seas, and healthy ecosystems, etc. Eco-city and green community can provide future generations with a stock of natural and social capital which can exceed our own inheritance. Actually, cities and communities are the most appropriate places to implement sustainable development. At the same time, no place can receive more potent and beneficial rewards from the eco-campaigns than our cities and communities. Some researchers have proposed urban ecosystem ideas to treat urban development issues. Since the core of eco-city and green community is a healthy and ecologically friendly society, cities themselves must be viewed as integrated ecological systems that are also connected with regional and global ecosystems. The idea of eco-city and green community should inspire urban planners and policy makers to design urban related policies. It provides a fundamental approach to change city development patterns and thus eventually change human behaviors through collaborative efforts of governments and the private sectors. While governments implement the concepts of eco-city and green community, private developers, who build for purely commercial returns, can change their behaviors from pure commercial commitment to more responsible environmental and social commitment to citizens and the quality of life. Berke (2008) has pointed out that a collaborative, comprehensive, systematic approach is needed to develop eco-city and green community campaigns.

OBJECTIVE

This book's objective is to present a comprehensive, theoretical, practical, and adaptive approach to understanding the issues involved in eco-city planning and green community development. It builds on recent advances in urban theories, environmental science, architectural design, engineering, and geospatial information technologies to provide readers with the scientific foundation needed to understand the major visionary ideas about new urban forms. This book provides a basis of knowledge in planning theory and natural science and a major review of urban forms that have evolved over the past century; its primary emphasis is to describe and explain emerging approaches, methods, and techniques for eco-city and green community planning, design, policies, and technologies. This book responds to the key questions outlined at the beginning of this introduction: What are the theoretical foundations and historical views of eco-city and green community? What is a green and eco-friendly urban form? How can we find the appropriate approaches to build eco-city and green community? What international experiences and lessons we can learn from?

BOOK STRUCTURE

Under the framework of eco-city and green community, this book is organized in individual chapters by authors. There are five major sections in this book.

Section I. The "Introduction" highlights the major problems in current urban development and urban planning and explains the objective of the book and the book structure.

Section II, "Evolution of Eco-city and Green Community" reviews the history and theories of urban planning and examines how major visionary planning theories about urban development have influenced the evolution of eco-city. This part builds a solid theoretical foundation and historical context for emerging eco-city and green communities. It also provides an up-to-date discussion of current practices with a vision of new urbanism and post-modern land use planning patterns.

Section III, "The Approaches to Build Eco-city and Green Community." This part introduces the broad concepts of urban and land use planning, and describes practical planning approaches and explains how to incorporate the core concepts of sustainable development into urban planning and community development. A comprehensive review of the methods is a guide for developing eco-cities and green communities. The major methods include livability and walkability techniques, transportation planning, environmental planning, renewable energy policies, collective community design skills, planning support systems, and public participation.

Section IV, "Models of Eco-City and Green Community." This section provides real cases and comprehensive reviews of major eco-cities and green communities. It explores the international context of urban planning and proposes a model to assess future conditions, evaluate policy choices, create visions, and compare scenarios. This section will identify eco-cities and green communities and analyze their experiences and policy implications. We will illustrate how what has been learned from sustainable urban form models and qualities can be applied to an existing metropolis. Finally, some conclusions are drawn about how the findings have helped develop an understanding of what sustainable form is and whether it can be achieved.

Section V, the "Conclusion" summarizes the major ideas from this book and provides suggestions for future research and practice of eco-city and green community.

REFERENCES

Berke, P. R. (2008). The evolution of green community planning, scholarship, and practice. *Journal of the American Planning Association,* 74(4): 393-407.

Berke, P. R, Conroy, M. M. (2000). Are We Planning for Sustainable Development?: An Evaluation of 30 Comprehensive Plans. *Journal of the American Planning Association,* 66(1): 21-33.

Girardet, H.t (2004). *Cities people planet*, Wiley-Academy.

World Commission on Environment and Development. (1987). *Our common future*, Oxford University Press, Oxford; New York.

SECTION II: EVOLUTION OF ECO-CITY AND GREEN COMMUNITY

In: Eco-City and Green Community
Editor: Zhenghong Tang
ISBN: 978-1-60876-811-0
© 2010 Nova Science Publishers, Inc.

Chapter 2

THE HISTORY AND EVOLUTION OF ECO-CITY AND GREEN COMMUNITY

Zhenghong Tang and Ting Wei

Community and Regional Planning Program
University of Nebraska, Lincoln, NE, USA

This section illustrates the evolution of eco-city and green community. It first provides an overview of major planning theories and planning history to highlight research relating to the concept of eco-city and green community and present the definition of eco-city and its derivation. Although there is not a concrete definition of the terms "eco-city" or "green community", it still represents the goal – a direction for sustainable city development. The section then explores the historical development of the eco-city and green community. Major planning theories and historical achievements have laid the groundwork for sustainable urban form models. Understanding particular historical themes – which continue to be echoed today – is important in order to understand how cities and communities can become more sustainable in the future. Based on the historical overview and theoretical discussion, we have also reviewed urban models that can provide insights for eco-city and green community development. It continues to question the proper eco-city models, compares different physical urban forms models and classifies the qualities of the eco-city system model. This section offers insights into physical models which may be more sustainable for urban ecology, land conservation, and transport. In so doing, the linear metabolism city model, the compact city model, the macro-structure of alternative city models, and dense city models are explored. After comparing and contrasting these models, the dense city model was determined to be the appropriate model in many aspects. The concept of eco-city and green community dispels the widely held view that sustainable urban development is a key step for achieving long-term global sustainability.

OVERVIEW OF PLANNING HISTORY

Urban planning involves the consideration of decisions before choosing among alternatives. The functions of urban planning include: 1) improved efficiency of land use, 2) balancing public and private interests, 3) providing a wide range of choices, and 4) promoting public participation in decision making. Planning is used to anticipate the future and can also be perceived as an agent of development. The concept of eco-city and green community is a part of planning history. Berke (2008) summarized the major development steps of green communities as planning ideals: harmony with natural systems, human health, spiritual well being and renewal, livable built environments, and fair-share community.

Early planning theories emphasized practical problem solving; however, planning has evidence an evolutionary nature since the 1900s. In the late 19th century, urban planning focused primarily on identifying urban problems originating from the industrial revolution and the growth of urban corporations. Urban decision makers dealt with unwanted urbanization effects such as fire, disease, pollution, etc. Pre-modern planning focused on urban design and street systems.

A brief summary tracks the main focus of early planning. The Philadelphia plan designed the grid system and neighborhood parks; the Savannah plan developed a ward park system. In the 1790's, Washington D.C. provided a model plan for the whole city. The most successful model plan was Central Park in New York City -- the first major purchase of parkland in a large city. It has now been recognized as a milestone in the eco-city and green community movement.

Early in the 1900s, urban planning moved to the institutionalization of planning. Significant progress in this period (1900-1945) involved planning toward institutionalization, standardization, professionalization and recognition of planning together with the rise of national and regional planning efforts. In 1907, Harvard University offered the first professional planning course in the United States. At that time, many cities had moved into professional planning. Washington D.C. created the first planning association while Wisconsin established state legislation to permit cities to develop urban planning. The City of Chicago created the first regional plan and Los Angeles implemented the first land use zoning ordinance. In 1910 in Germany a master plan was created in the Greater Berlin area. Since urban areas have become important political and economic centers at the national level, a series of parliamentary statutes have been enacted to regulate planning systems. At the first half of the 20th century, many American cities made significant progress in urban planning. Massachusetts established state mandates requiring local governments to provide planning boards. Los Angeles County established a planning board in the 1920s and New York City created a comprehensive zoning ordinance. There were increasing concerns about social welfarism, urban health, and housing issues. For example, a comprehensive plan by the City of Cincinnati was based on the welfare of the whole city. With increased industrialization, public health became a serious problem for urban planners because large numbers of people were living in unhealthy environments. In addition, land demands rapidly increased land values, particularly for land with good environmental quality. In the 1930-1940s, focus on physical planning was a significant characteristic of the urban planning movement.

After World War II, the emergence of new economic models changed the role of planners. Post war era planning was developed for standardization, crisis, and diversification.

Planners played an important role in social change rather than just in urban design when social inequality and poverty were considered the planners' scope. At the same time, welfare capitalism changed the focus of planning to enterprise and economic development, thus planning decisions have increasingly been linked to political and economic decisions especially when developers exert important influence on urban development. High land values caused more social inequality and environmental problems. Fortunately, a positive trend in modern planning has increased public participation in planning decisions with a bottom-up planning model required in many planning decisions. Since the 1980s, the concept of sustainable development has become a core value in planning decisions.

The brief history of urban planning paints an evolutionary picture of the planning system showing that urban planning has experienced significant change in scope and focus. Although there are still debates about the role of planning, it has been widely accepted by planners that sustainable development is a core value in urban planning.

THE ROLE OF PLANNING THEORY

Planning theory is the arena where the field of planning reassesses itself. Since planning decisions deal with complex scientific and social issues, it is necessary to clarify the role of planners. Planning theory gives a basis for the kind of data planners should collect and guidelines for making decisions as well as encourages planners to explore many fundamental planning questions. Planners must know who the plans work for, the focus of the plans and whether there is equitable distribution of resources.

Rational planning theory is the basis of most public planning. In general, rational planning includes 1) identifying the planning problem, 2) forming the goal of planning, 3) collecting planning data, 4) identifying planning approaches, 4) assessing alternative scenarios, 5) selecting preferred alternatives, 6) plan implementation, and 7) monitoring and updating plans. Rational planning first clarifies the planning mission, role, and responsibilities, then delineates intended purpose and determines plan framework and format. In rational planning planners determine methodologies, participants, coordination and evaluation procedures. Through rational analysis, planners identify needs and priorities, and eventually develop implementation, monitoring, and updating strategies. The rational planning model provides justification for planning which is the birth of scientific planning. A rational planning model can achieve objectives such as order, efficiency, and cost effectiveness. The success of rational planning depends on a high level of knowledge and technical skill. However, rational planning has its weaknesses -- critics point out that traditional rational planning may have limited time and financial support to collect data. The data itself may be limited and imperfect and thus be difficult to use for rational planning decisions. In addition, it is extremely difficult to integrate societal values into a scientific planning process. Moreover, urban planning itself is an open, rapidly changing system with many unforeseeable consequences, and rational planning cannot ensure consistency within a dynamic system. Although rational planning theory has been increasingly criticized, it is the most dominate theory in planning.

Incremental planning theory was developed in the 1960s as an alternative theory that accepts most shortcomings of rational planning. Incremental planning believes that planning is incremental, disjointed, dynamic, and serial. It supports the comprehensive approach, but pays more attention to the incremental approach.

Communication and collaboration theory provides an opportunity to reformulate traditional rational planning. It offers more opportunities for stakeholders and citizens to participate in the planning process to promote higher visibility and transparency of decision making. Communication and collaboration promote mutual understanding of planning issues. A significant shift has occurred from a rational planning paradigm to communicative planning since the 1990s.

THE ROLE OF PLANNERS

Planners must address numerous challenges and reconcile the goals of economic development, social justice and environmental protection while striking a balance between their expertise and citizen input. In practice, the planner's role has been extended beyond technical activities to address social, economic and environmental challenges. It is necessary to have technical support for effective decision making.

However, the role of planners is subject to political constraints. The planner's decisions may be influenced by developers, consumers, and other power groups since their decisions frequently rely on either developers' investments or political will.

It is important for planners to involve public interests in planning decisions in order to provide equal protection, equal opportunity, environmental justice, and equal social responsibilities; thus a major challenge for planners is to define the boundaries between public and private interests. Planners can facilitate public involvement, encourage citizen participation as well as negotiate and mediate environmental conflicts.

DEFINITION OF ECO-CITY OR GREEN COMMUNITY

The term "eco-city" is comparatively new, and difficult to conceptualize. The concept of eco-city or green community does not immediately suggest one particular urban form. Because urban situations vary, it is difficult to suggest a preference for certain densities, decentralized or centralized development, or the exact size of settlements. The definitions of eco-city or green community are discussed in works such as *Towards Sustainable Communities* (Roseland 1992), *Towards an Eco-city* (Engwicht 1992) *and Building Sustainable Communities* (Nozick 1992), but there is no consensus on a definition because the implications of eco-city and green community are still developing, thus policy makers and researchers are still not sure about the scope. Nevertheless, all the discussions about eco-city or green community (or sustainable communities, or sustainable cities, or eco-communities), represent one common goal — an ecologically friendly development of communities. This agreement can be used as a benchmark to guide the development of eco-city and green community. This core value can evaluate the quality of community and rethink the sustainable urban forms that our cities or communities should have. The value of eco-city and green community provides a useful reference point for solving current urban problems. Therefore, at present, it is safe to say that there is no single accepted definition of "eco-city" or "green community," but there are some common principles. Inherent in much of the literature is the recognition that eco-city and green community are considered a part of local action for sustainable development. Thus, eco-city and green community must define sustainability from a local perspective. Of course, the immediate challenge is how to incorporate regional or global environmental values into local eco-city and green community development. It is also a problem to build local social equity within a framework of global sustainability (Roseland, 1997; Tang, 2009). In summary, an eco-city or green community should be environmentally friendly, socially equal and self-sufficient in energy, water and food production. Besides changes in the physical environment, it is important to consider the characteristics of production modes, consumption behavior and decision instruments. The campaign for eco-city and green community should encompass a comprehensive, systematic view and collaboration in urban planning, transportation planning, public health, housing policy, energy policy and technologies, natural resources management, and social justice. The label eco-city or green community creates an innovative image that incorporates the natural and social environments. It emphasizes significant changes in a city's urban planning to create a healthy community. It also asks for changes in economic development to ensure that the new image is successfully merchandized.

U.S. EPA (2009) proposed key points to describe a green community to help understand the definition of eco-city and green community. EPA has established three major goals for the environment, economy, and society. In the environmental aspect, a green community must comply with major environmental regulations and focus on waste reduction, pollution prevention, and natural resources conservation through sustainable land use. In the economic aspect, a green community should promote diverse, locally-owned or operated sustainable businesses, adequate affordable housing, mixed-use residential areas, and economic equity. In the social aspect, a green community needs to promote active public participation and mutual communication, incorporate local values and identity, create safe and friendly neighborhoods, provide efficient infrastructure, and promote equitable and effective educational and health systems. The above guidelines from the EPA provide the fundamental principles to build eco-cities and green communities. Cities that engage in ecologic or green planning should integrate their environmental, economic, and social goals into broader sustainable development to provide a balanced urban development pattern.

EPA (2009) also provides basic implementation strategies to apply the principles of eco-city and green communities. According to EPA suggestions, a community-based comprehensive assessment is the first step toward a green community where a community delineates the planning area boundary, identifies community core values, and provides a detailed inventory of natural and built environments. Based on basic factual information, the community can conduct an evaluation of its economic conditions, infrastructure capacity and effectiveness. If possible, the community should analyze the linkages among social, economic and environmental issues in the development process and make comparisons with peer-communities or cities. The second step is to conduct a trend analysis. A community-based assessment allows communities to anticipate the development patterns they need and visualize multiple sceneries. The third step is to set specific goals within specific time frames by linking the community vision with goals that lead to the future of an ecologically friendly environment. By incorporating public participation, communities can brainstorm ideas that actually reflect citizens' value and interests. The fourth step is to find solutions and strategies by recommending strategies, assigning responsibilities, setting priority actions, developing detailed implementation procedures, and committing financial sources.

This study further conceptualizes the definitions for eco-city and green community as: 1) An eco-city and green community protects and preserves the natural environment. Undisturbed sensitive natural resources are protected. 2) An eco-city and green community supports local agriculture and local products to reduce the ecological footprint at the community level. 3) An eco-city and green community encourages a mixed-use, compact, clustered urban land use development pattern. 4) An eco-city and green community develops transit-oriented, pedestrian-friendly community by promoting walking, bicycling, and public transit use. 5) An eco-city and green community develops efficient transport and communication systems to change traditional human behaviors by using new technologies to reduce costs, improve efficiency, and save energy. 6) An eco-city and green community maximizes renewable energy and adopts energy conservation strategies. 7) An eco-city and green community adopts zero-waste programs to recycle materials. 8) An eco-city and green community makes decision makers aware of the concept of ecosystem and sustainable development. 9) An eco-city and green community promotes stakeholder involvement and community-based efforts to improve community quality. 10) An eco-city and green community encourages public participation to support development campaigns.

ORIGINS OF THE ECOLOGICAL CITY AND GREEN COMMUNITY

Although the term "eco-city" looks relatively new, the concept was actually born in the late 19th and early 20th centuries. At that time, urban planners and managers sought to draw attention to the deterioration of urban conditions, particularly public health and realized the necessity for alternative living environments. One of the most influential writers who addressed this problem was Ebenezer Howard.

Howard (1898) looked for a balance between core city and country life which has become the central task for creating more sustainable communities in current urban planning. Of course, the emphasis of his original idea has shifted. In the 19th century, cities were not extremely dense, but urban living conditions in many cities were horrible lacking adequate clean water, basic sanitation, and decent housing. At the present time most U.S. cities have relatively low-density with car-dependent suburbs with high quality infrastructure and housing. So the question now is how to rethink the balance of core city and country life which was the main focus of Howard's idea in the 19th century. Planners must create new forms of garden city to avoid the problems of overcrowded industrial cities and rapid suburban sprawl with low density building.

Lewis Mumford (1938) and Ian L. McHarg are two key persons in eco-city development who seek different implementations of the eco-city idea. Although Mumford is a writer rather than a planner, just like Howard he played a central role as America popularized the garden city idea. His solution for the problems of the overcrowded industrial city advocates the decentralization of population to achieve a better balance of city and countryside. McHarg's book *Design with Nature* (1969) is the milestone in modern urban planning theory with ideas that have influenced current urban planning theory and practice. His idea of designing with nature has played a crucial role in bringing environmental and urban planning concerns together in the mid-20th century. Later books have emphasized the importance of ecological principles in urban development - *Silent Spring* (Carson, 1962) called attention to the dangers of pesticides and other toxic chemicals. Additionally, Barry Commoner's book *The Closing Circle* (1971), warned of the impacts of pollution and resource consumption in a technological society. All of these books from the 1960s and 1970s helped catalyze the modern environmental movement.

Jane Jacobs is also a significant 20th century writer in the urban planning field whose specific emphasis is on pedestrian-oriented urban forms. In *The Death and Life of Great American Cities* (1961), Jacobs criticized the modernist planning models that have destroyed many existing inner-city communities. She advocates a dense and mixed-use urban model and encourages pedestrian use of the street, neighborhood contacts, and a thriving local economy of small businesses. One of her major goals is to preserve the uniqueness inherent in communities upholding redundancy and vibrancy of neighborhoods against order and efficiency. She described in detail what makes dense urban neighborhoods work, and how modern city-building practices undermine many neighborhood qualities. Her ideas have served as an inspiration to many later urban activists including the "New Urbanists."

The Club of Rome and Donella H. Meadows, Dennis L. Meadows, Jørgen Randers, and William W. Behrens III are the most influential planning researchers in planning theory. Their book *The Limits to Growth* discussed the consequences of a rapidly growing world population with limited resources. They determined that our resources are limited after they analyzed growth issues from five critical aspects -- world population, industrialization, pollution, food production and resource depletion -- and concluded that the future growth of human society on our planet would be limited. They believe that resources may be depleted within a hundred years which will lead to a steep decline in global population and industrial capacity thus resource depletion, pollution (including carbon dioxide concentration), loss of arable land, and declining food production would converge to halt world growth. However, they suggested that it would be possible to alter these negative trends *"by establishing a condition of ecological and economic stability that is sustainable far into the future"* (Meadows & Forrester, 1974, P84). This work first used computer models to analyze the human future to answer critical questions such as whether growing human population and resource consumption were sustainable. However, this book and its opinions became dramatically controversial as soon as it was published. Its research was criticized as irresponsible nonsense because insufficient evidence was used to support the variables used; critics also cited a lack of strict scientific procedures to justify the conclusions. More important, this book did not reflect the dynamics of growth even if resources may be finite.

Richard Register is a key researcher who has promoted the research and practice of eco-city. In 1975, he and his colleagues founded Urban Ecology in Berkeley, California, as a nonprofit organization to "rebuild cities in balance with nature." Register's publication *Eco-city Berkeley* (1987) and the institutional journal *The Urban Ecologist* created a new momentum in the study of eco-city and green community. They proposed the concept of urban ecology as a subfield of ecology to analyze the interaction of ecology with the human environment in urban or urbanizing settings. They paid special attention to the effects of urban development patterns on ecological conditions and suggested that urban planners use urban design strategies and new building materials to promote a healthy, biodiverse, urban ecosystem. David Engwicht (1992) is another important researcher in eco-city and green community who pointed the way towards "eco-cities" where people can move via foot, bicycle, and mass transit thereby improving existing transportation systems and changing daily behaviors to reach the goal of eco-city.

In 1990, the first International Eco-city Conference was held in Berkeley which discussed urban ecological problems and possible approaches for shaping cities upon ecological principles. After the second International Eco-city Conference (held in Adelaide, Australia, 1992) and the third Conference (held in Yoff, Senegal, 1996), Urban Ecology set forth ten principles for creating ecological cities.

1) revise land use priorities to create compact, diverse, green, safe, pleasant, and vital mixed-use communities near transit nodes and other transportation facilities;
2) revise transportation priorities to favor foot, bicycle, cart, and transit over motorcars, and to emphasize "access by proximity";
3) restore damaged urban environments, especially creeks, shore lines, ridgelines, and wetlands;
4) create decent, affordable, safe, convenient, and racially and economically mixed housing;
5) nurture social justice and create improved opportunities for women, people of color and the disabled;
6) support local agriculture, urban greening projects, and community gardening;
7) promote recycling, innovative appropriate technology, and resource conservation while reducing pollution and hazardous wastes;
8) working with businesses to support ecologically sound economic activity while discouraging pollution, waste, and the use and production of hazardous materials;
9) promote voluntary simplicity and discourage excessive consumption of material goods;
10) increase awareness of the local environment and bioregion through activist and educational projects that increase public awareness of ecological sustainability issues.

The above ten principles are theoretical guidelines for the practice of eco-city and green community. The international conferences have become regular forums to discuss important theories and problems in eco-city and green community. Those annual conferences have significantly promoted the development of eco-city and green community.

MODELS OF THE ECO-CITY OR GREEN COMMUNITY

The theoretical models of eco-city or green community have provided many meaningful insights for urban designers, planners, and decision makers. A number of cities and numerous communities around the world have embarked on the long road towards sustainable eco-city or green community to solve existing urban problems. The models for eco-city and green community provide useful theoretical guidelines for possible solutions. This section mainly discusses two general physical models: the linear metabolism city model and the compact city model which have been used in many cities. The compact city model has been widely accepted as the core of sustainable urban research; however, there is inadequate description of this model. Therefore, many urban planners have developed their own urban models based on the compact city model or other related models. These alternative urban models include eco-village, eco-neighborhood and multiple models. All of those alternative urban models are macro-level urban structure models which provide new ideas for the physical development of urban forms. The dense city model is increasingly accepted as the solution for sustainable urban forms. The following sections discuss the major urban models.

Linear Metabolism City Model

The term "metabolism" can be defined as *"the sum of all the biological, chemical and physical processes that occur within an organism or ecosystem to enable it to exist indefinitely"* (Girardet, 2004). The metabolism city model describes the flow of materials and energy within cities and provides a systematic overview to analyze all of the activities of a city from the materials and energy flow aspects. This model tries to find systematic solutions for complex urban problems by calculating resource consumption. The keystone of this model lies in cities aiming at a circular 'metabolism,' where consumption is reduced by implementing efficiencies and maximizing re-use of resources (Rogers, 1997). The linear metabolism city model demonstrates how the processes of urban production, consumption and disposal undermine the overall ecological viability of urban systems. Urban planners can use this model to analyze whether or not resources are used in a sustainable way. However, the simple linear metabolism city model does not represent real city activities. Urban planners actually need to adopt the circular metabolic systems rather than the linear metabolism system to assure their own long-term viability and that of the rural environments on whose viability they depend. It is difficult to linearly link ecological footprint reduction with the flow of materials and energy. For instance, community health and livability cannot be calculated as a direct factor in the linear model. The outputs from a city or a community will also need to be inputs into the urban production system with routine recycling of paper, materials, plastic and glass, and the conversion of organic materials, including sewage, into compost, returning plant nutrients back to farmland to maintain soil health. Thus, the city itself is much more complex than the linear metabolism city model.

The 'Compact City' Model

The Commission of European Communities (CEC) Green Paper published in 1990 clearly calls for a return to the compact city. In the past two decades, numerous researchers have addressed the compact city model which is becoming the most popular city model for urban studies. While urban sprawl has become the primary form of urban development in many countries it has received increasing criticism for its negative environmental, economic and social impacts; thus many researchers are seeking alternative models to put cities on the right track by reducing human impact on the environment. Although there are still many debates about whether the compact city model can be considered as a perfect sustainable urban form, it has been widely incorporated into many urban redevelopment plans. It is now widely accepted, particularly in land use planning policy, that the most effective solution to achieving sustainable urban form is implementation of the compact city model, that is, encouraging the high-density, mixed-use urban form. For example, mixed-use and compact development policies have been mandated for California's local planning systems. Supporters of the compact city model argue that the compact city has environmental and energy advantages and social benefits (e.g. CEC, 1990, Jacobs, 1961, Newman and Kenworthy, 1989). This group has proposed some core principles for the compact city model, including: 1) strong supportive communities, 2) a high-quality living environment, 3) good access to other places, and 4) protection of natural environment.

On the contrary, many urban researchers argue against the compact city believing that compact development itself cannot lead to a significant change in urban development patterns because it does not reflect the hard reality of economic demands, environmental conflicts and social context. It is unsure whether the compact city model can change the existing low-density, single-use, automobile dependent type of development. Since the traditional urban model has dominated urban planning patterns for more than fifty years, it has already created a series of regulatory, economic and cultural problems. The compact city model may partially address some problems, but not enough to change cultural and regulatory or economic factors since it has ignored the causes and effects of city decentralization and hasn't found the right solutions for decentralization (Thomas and Cousins, 1996, P56). Some researchers have asked for a reexamination of the compact city model because it is still not clear what the structure and form of such a city might actually be. The Commission of the European Communities' Green Paper clearly calls for a return to the compact city, influenced by the fact that many historic European towns and cities have densely developed cores which are seen as ideal places to live and work. Such places have high population densities which encourage a social mix and interaction which are the major characteristic of traditional cities. However, most American cities do not have this historical context; thus the compact city model is not likely to be accepted soon. For example, encouraging public transport and facilities for walking and cycling is an advantage of the compact city which can be improved by modern industrial city conditions. Therefore, the compact city model also has its limitations. Following the concept of compact city, Hildebrand Frey (1999) has expounded his own model for the sustainable city form and development.

The Macro-structure of Alternative City Models

While researchers have criticized the linear metabolism city model and compact city model, they have turned their attention to alternative macro-structure models which illustrate the different development patterns for a city region or metropolis. This approach aims to determine how the individual models compare in terms of overall dimensions and total area required and how they respond to sustainability. The following sections will explain the submodels.

The Core City (Lynch, 1984) model emphasizes a city core with a very high density and an intense peak of activities at the center which provides all city functions and is designed to be concentrated into one continuous body in a small area. The advantages of the core city model are reduced distances for materials, goods, and services. It will provide access to facilities and the countryside if population and city size remain relatively small. The smaller core cities will be organized as part of a larger core city which is composed of a number of such cities. The linkage between small core cities and larger core cities depends on public transport to provide services and facilities beyond these provided by the smaller individual cities.

The Star City model (or urban star model) is another of Lynch's city models. This model has a single dominant center as well as its own transportation system comprised of public transit and main vehicular traffic routes. The dominant center model includes secondary centers with high to medium density. Public transport routes link all the secondary centers with the dominant center with transport stops located in the sub-centers of the city. According to Lynch's ideas, the star city model would be more extensive in its area and size when compared with the core city model. Lynch thought that the star city model should include a certain amount of open spaces increasing with the growing city. The advantage of this model is that it allows the city to grow farther outward than the core city model. Of course, it is a challenge to expand the city and avoid congestion at the center and avoid overloading radial transport lines across the city.

The Satellites City model (Howard, 1898) has been widely applied in many cities around the world. This conceptual model includes a central city surrounded by a set of satellite towns or communities with some distance between the central city and surrounding towns. Also, according to Lynch's concept, the satellite towns' optimum population should between 25,000 and 250,000. The significant difference between the satellite towns and sprawling suburbs is the independence of the communities. Satellite towns generally have independent city functions in employment bases, city transport, and cultural centers as well as an independent municipal government. Satellite towns do not just provide city functions as bedroom communities - the central city and the satellite towns play different roles in city services. This model does not encourage continuous sprawling development in urban areas, but in separate communities. The limitations of this model indicate that core cities are generally thought to be less efficient with a lower quality of environmental and living conditions. The buffer areas between the city and satellite towns are the key to maintaining city quality. This model has been used in many cities as listed in Table 1a.

The Galaxy of Settlement (Lynch, 1984) describes an urban form based on existing old centers and subcenters. It encourages continuous decentralization and divides the old centers (or subcenters) into small units which can grow as relatively dense central cores linked by a network of communication and transport lines. While the centers and small units have relatively higher densities, they are separated by low-density suburban areas or open land. This idea is consistent with the satellite town model and star city model. Although the galaxy of settlement is an ideal model which is little used in practice, it has actually influenced current planning theories, for example, Calthorpe's transit-oriented development pattern. Also, Duany and Plater-Zyberk's traditional neighborhood development (TNDs) pattern uses some of the ideas in small towns. The Galaxy of Settlement model provides a modern vision for traditional towns and further develops traditional towns in a regional context because it believes that the traditional towns should not just be self-sufficient. This model supports the development of a central metropolitan or regional city with links to a number of small towns.

The Linear City model, as a dynamic alternative to the more static core city model, consists of a number of parallel cities which have different specialized functions. The linear city is linked with a continuous transport line or parallel series of lines so it encourages developing a public transportation system for commuting. It may include a segregated zone for transport lines, a zone for industrial and communal enterprises, a green belt or buffer zone with transport lines, a major residential zone, an open space zone, and an agricultural zone to provide local food. The major urban activities are arranged along and on either side of the lines. A relatively dense compact strip of development will be formed along the transport lines, with other places having low dense development patterns. There is no unique central core for the linear city model, however, the whole city will become a compact model. However, the linear city model may be economically viable only in the form of cross-city links (Minnery, 1992). Currently there are relatively few cities which have implemented this model.

Table 1a. Satellite towns of the metropolitan areas

Satellite towns/cities	Metropolitan areas	Satellite towns/cities	Metropolitan areas
Akron	Cleveland, Ohio	Atlantic City	Philadelphia, Pennsylvania
Ann Arbor	Detroit, Michigan	Aurora	Chicago, Illinois
Boulder	Denver, Colorado	Elgin	Chicago, Illinois
Escondido	San Diego, California	Flint	Detroit, Michigan
Fort Lauderdale	Miami, Florida	Fort Worth	Dallas, Texas
Frederick	Baltimore, Maryland	Joliet	Chicago, Illinois
Long Beach	Los Angeles, California	Newark, New Jersey	New York City, New York
New Haven, Connecticut	New York City, New York	Oceanside	San Diego, California
Pasadena	Los Angeles, California	Santa Rosa, California	San Francisco, California
Tacoma	Seattle, Washington	Worcester	Boston, Massachusetts
Gurgaon, Haryana	Delhi, India	Navi Mumbai	Mumbai, Malaysia
Colchester	London, United Kingdom	Luton	London, United Kingdom
Colchester	London, United Kingdom	Reading	London, United Kingdom
Luton	London, United Kingdom	Changping, Beijing	Beijing, China

The regional city model or polycentric net (Lynch, 1984) emerges as a new urban pattern. In the 20th century, many cities developed as networked polycentric mega-city-regions which are characterized by a cluster of cities, towns, and communities. These cities are physically separated but intensively networked together. In the United States, the major mega-city-regions include: Boston–New York City–Philadelphia–Baltimore–Washington, Los Angeles–San Diego-Tijuana, and San Jose–San Francisco–Oakland – Sacramento. In China, there are some mega regions in the Yangtze River Delta Metropolitan Area, Pearl River Delta, and the Beijing-Tianjin corridor. These corridors resemble a complex circulation system which grows in any direction. In general, the whole region has a wide range of different densities. An efficient transportation system is the key to reduce the peaks of traffic and linear concentrations in the regional city model or the polycentric net. Higher density development will be located in the nodes of transport lines and linear lines and lower density will be planned in other areas. This model combines the core cities with the linear cities and other urban stars in a mixed regional planning pattern. It shows the characteristics, structures and form of regional development patterns in which a central metropolitan core is not always necessary (Minnery, 1992).

All in all, the fact that city models all score reasonably well under different weighting may mean that all of them may work well on their own or in combination to become a sustainable city region.

The 'Dense City' Model

The concept of "Dense City" city was proposed by Richard Rogers in 1997. Dense city emphasizes the overlapping of urban economic and social activities which bring social opportunities and ecological benefits through spatial density and social diversity. The dense city model can reduce necessary urban activities and therefore mitigate inconvenient impacts related to urban size. The dense city model can build a city at low cost since it can significantly save the investment spent on transportation including roads, parking lots, and urban infrastructure. It can also reduce costs for good housing and transportation as a result of short travel distances. The dense city model can increase energy efficiency, reduce travel time and distance, reduce accident rates, avoid traffic congestion, consume fewer resources, produce less pollution, and save money for residents. Therefore, this model can create new public space, protect natural environment, and improve the quality of air while reducing destructive urban sprawl and protecting the countryside, the environment, and the ecosystem in the long term. It benefits quality of the countryside by protecting land from the encroachment of urban development (Rogers, 1997). Additionally, the dense city size provides more flexibility for adjusting city development patterns.

A dense city includes important characteristics such as well-designed urban structures, compact development patterns, and good connections and accessibility. The dense city model has been developed as a flexible structure with parts related to the whole. The dense city model encompasses decentralization which promotes the effective development of local centers. It also emphasizes the importance of public space in the dense city model since a clear articulation of public space can provide accessible connections to different communities and link activities among schools, work-places, and basic social nodes.

However, there are many critical problems in the dense city model. For example, deciding the appropriate possible density is subject to a number of factors therefore it is difficult to transfer the dense city model to citizen's day-to-day lifestyles. The dense city model must strike a balance between city compactness and social equity, urban periphery, social exclusion, and community security issues. In addition, the dense city model must be frequently updated with new information, especially social issues such as changes in family size, housing needs, education levels, etc.; thus, the dense city model must incorporate those changes into new urban decisions. Since cities are dynamic organisms as complex as society itself, city planners need to react swiftly to new changes and new information. The dense city model also brings uncertainties for urban vibrancy. A city with dense development patterns needs to meet a series of needs to achieve sustainable urban development. We agree with Rogers's (1997, P167) description of the dense city as a city with many facets: 1) multicentered, 2) overlapping activities, 3) ecologically friendly, 4) easy-commuting, 5) social equality 6) open space, and 7) beauty if possible.

COMPONENTS OF THE ECO-CITY AND GREEN COMMUNITY

The following section describes the components of the eco-city and green community system. While researchers have discernible differences in their focus, methods, models, and strategies regarding eco-city or green community, the main theme can be conceptualized as different components. The concept of eco-city and green community does not stand alone but is incorporated within a complex social context. Components of the eco-city and green community possess a complex array of relevant variations in regard to social, environmental, and economic aspects. This section discusses several components of eco-city and green community.

The first component is *land use and urban design*. In any city, the daily life and vitality of urban place is the core of urban development (Jacobs, 1961). Since the 1960s, a number of urban researchers (e.g. Jacobs, Lynch, William H. Whyte, Clare Cooper Marcus, and Danish designer Jan Gehl) have related urban design to urban living quality. Their research studied how people actually experience and use urban environments. Forty years later, environmental design is emerging as an important discipline in urban planning and landscape architecture devoted to improving built environments. Urban design is an effective approach to build livable cities and create sustainable urban life. A city with advanced ideas of urban design can create high density and mixed land use urban forms to protect natural resources and encourage compactness and social interaction. An eco-city or green community should maintain the long-term goals of high-efficiency with a small-footprint urban design. Although the traditional sprawling model attracts property developers and individual homeowners in the short term, it brings a multitude of problems such as traffic congestion, loss of productive land, and increased environmental pollution which degrade the quality of life. Not only can land use planning policy and urban design strategies save land from development, but, more important, they can provide alternative solutions for urban sustainability. One important aspects of land use planning and urban design is the protection of urban open spaces which are a vital part of the urban landscape. Urban open spaces provide precious natural areas for outdoor activities and a venue for a range of social activities directly related to residents' health. Land use policies and urban design strategies should foster a lightly populated urban form and promote mixed-use of shops, dwellings, offices and public spaces. Land use planning and urban design strategies should improve the design of street patterns which should not be based on a grid but on street systems with more efficient, circuitous, branching patterns.

The second component is the *transportation system* – the center of debates about sustainable urban forms. In traditional planning models, the majority of urban development investment was directed toward the expansion and new construction of roads and transportation systems to meet fast growing transit needs. Although many cities are willing to move towards a compact urban form, it cannot be easily implemented in practice, thus there is still a gap between the new compact urban form and daily human travel patterns. *Cities and Automobile Dependence* (1989), authored by Peter Newman and Jeffrey Kenworthy, analyzed the relation between urban density and petroleum consumption and showed that the range of urban densities worldwide are strongly correlated with decreased resources. In their later book *Sustainability and Cities* (1999), they provide strategies for moving away from automobile dependence. Simmonds and Coombe (William, et. al., 2004) then developed a transport model to test various urban forms and transport scenarios. They found that compact city strategies do not necessarily have a direct impact on total demand and car use. Their research also pointed out that the compact city pattern does not necessarily indicate an efficient transport system. The reason identified by Simmonds and Coombe is that proximity has only a weak influence on travel choices. Recent research has called attention to wider strategies in transport systems that can be used to improve the urban environment such as significant changes in land use activity, urban infrastructure and services, and travel behaviors. For instance, some planners have actively promoted community awareness of transport systems and travel behaviors through local dialogues and by supporting local initiatives. Local planners have considered household transport costs as an issue for providing affordable transport and housing. The governmental role is to provide more transport options that citizens will use. It is unrealistic to give up private travel patterns completely, but behaviors can be modified if other options are sufficiently attractive.

It is important to further improve land use planning to support infill development over sprawl. An urban development model should provide an efficient framework for providing more housing and working opportunities in existing urban centers, and revitalize suburban centers. Active urban centers can reduce travel demand between work places and living places. Cities and communities should provide more frequent and better connected services to support a network city pattern that can significantly improve the efficiency of public transport. The potential road system should not only provide a conduit for motor vehicles, but should also meet different community needs for bicycles and pedestrians. Urban planners can reduce the amount of land used for transport systems by reducing the number of parking lots.

The third component is *urban ecology and restoration*. While urbanization is an important source of degradation of urban ecosystems, urban planning plays an important role in urban ecology protection and restoration. The urban environmental should not be treated as a settlement in a closed system, but should be viewed as a system of urban ecology. Some recent techniques have been used to monitor resource flow, select urban ecological indicators and performance criteria, and analyze the ecological footprint. Urban environment is considered particularly relevant to bioregions and river catchments (Tomatty, 1994). These studies have advanced the understanding of the complexity of the urban ecosystem. Urban settlements are not only physical systems, they are also human habitat. Urban ecosystems are highly related to geographical scale. While some urban development activities do not have obvious environmental impacts within small sections of a city, they may have a dominant environmental impact on urban ecosystems that are planetary in size. Historically, there are three categories of urban environmental addenda: "brown," "gray," and "green." At the beginning of industrialization, many western cities could be categorized into "brown" because of serious environmental health problems caused by industrial pollutants, accumulation of solid waste, overcrowded housing, low-quality community, and inadequate community services. After the cities solved the most serious direct environmental problems, they moved to "gray" which was indirectly linked to urban issues such as groundwater protection. The third is "green" related to urban environmental protection. Urban ecosystem restoration focuses on the effect of human impacts to the dynamics of indigenous ecosystems (Riley,

1998). Urban ecosystem restoration aims to establish a sustainable and healthy relationship between nature and humans. In the late 19th and early 20th centuries, the urban environmental movement focused on "conservation" or "preservation" of natural lands and species rather than on a more systematic approach to protect the environment. Since the 1980s, many urban studies have shifted to the restoration of the urban ecosystem as an urban sustainability agenda. Major urban restoration activities have included cleaning up contaminated lands (often known as brownfield sites), replanting native vegetation, and restoring water bodies, and cooperating with watershed management.

The fourth component is energy and material use. Urban planning provides an important approach to efficient energy and material use by improving the efficiency of energy and material use and helping minimize the amount of waste created while providing cost savings for city residents. Eco-friendly products are readily available to be used in urban planning while new construction techniques help cities build more environmentally friendly communities. The flow of energy and materials into cities is an important circle and a serious challenge to urban sustainability. If cities and communities recycle resources by reusing and re-manufacturing they can change the traditional materials management cycle and divert materials from their usual destination in landfills and incinerators. It is always the first law of conservation to reduce the consumption of energy and materials through appropriate land use policies. An eco-city or green community model needs to incorporate more renewable energy sources such as wind power, solar power, geothermal energy, biomass conversion, and co-generation.

The fifth component is environmental justice and social equity - items by far the least represented within urban development policy decisions. According to Elkin (1991), sustainable development does not involve just environmental conservation; it should embrace the need for equity. Elizabeth Burton (2004) has found that higher-density cities promote social equity. Urban environments are oversaturated with multiple sorts of pollutants in highly dense areas, thus public involvement is the most effective way to reach the goal of environment justice and social equity. Community policy makers and developers should be well versed in the principles of environmental justice and social equity and consider the cumulative environmental and social impacts of when implementing local projects.

The sixth component is economic development which can provide the sources and support for sustainable development. Unfortunately there are many unavoidable conflicts between economic development and urban environmental protection, e.g. long-term sustainable efforts may increase short-term costs. Traditional economic tools do not adequately incorporate externalities such as environmental pollution, resource depletion, loss of biodiversity, and degradation of the living environment. Also, traditional economic development patterns do not effectively take long-term costs and benefits into short-term account since traditional economic development models assume endless growth in material consumption even though it's unrealistic. Eco-city and green community models need to incorporate environmental and social factors, and seek new tools and techniques to balance social, environmental, and economic objectives.

The seventh component is green architecture and community. Green community needs to encourage residents' physical activities and build appropriate sidewalks, bicycle lanes, and interconnected greenways. This includes a comprehensive design so children can walk to school, adults can bike to work, and communities provide for more physical activities and social connections. Green community can minimize the amount of land needed for the built environment and thus provide opportunities for the preservation of wilderness areas. The principle of green community can reduce automobile travel and generated pollutants. Green building techniques are a current topic in architecture since builders realize that they can save resources such as energy, water, land, and raw materials through improved designs for buildings and operations. Since the first wave of modern green architecture emerged in the 1960s and 70s, solar energy technologies have been used for passive heating. Although renewable energy strategies are still optional in architectural design, energy conservation is now required by new building codes in many cities around the world. Since the 1990s, many pioneering firms have adopted new energy efficient strategies by using recycled building materials and efficient equipment in their construction. Since the 1990s and early 2000s, green building and green community have become the new philosophy for construction. Since LEED (Leader for Energy and Environmental Design) standards were first codified by the US Green Building Council in 1998 architects have begun to incorporate green features into their projects; however green building practices are still far from mainstream implementation.

No matter whether countries are rich or poor, there have been some planning successes and failures. First, many European cities have made successful efforts in historical preservation while maintaining economic vitality in their traditional city centers. Additionally, some European planning models have successfully linked economic regions with efficient public transport networks. Many European cities have also made significant achievements in environmental conservation and livable community design. However, many others have failed to control the growth of metropolitan areas. Second, American cities have created many successful ventures in open space and city parks such as New York City's Central park. Other cities have integrated air quality policies as a part of their urban planning decisions by implementing very strict pollutant emission standards to reduce car pollution. However, land segmentation and decentralized downtowns have been failures in many American cities. Public transportation systems have also failed in many American cities.

KEY ELEMENTS IN ECO-CITY AND GREEN COMMUNITY PLANNING

This study further illustrates the key elements for eco-city and green community planning. An ideal plan for an eco-city or green community should address the factual basis and provide appropriate strategies.

The factual basis section explains the basic environmental conditions of a city or a community and covers the key elements in a community's natural environment, built environment, and human health.

- *Local environmental setting:* An eco-city and green community plan should first describe a local jurisdiction's physical setting and include fundamental environmental characteristics.
- *Sphere of influence:* An eco-city and green community plan should not only address the environmental setting descriptions within its own political boundary, but also consider its potential service impacts on its entire planning area. This information can provide a benchmark for planners to consider regional environmental problems.

- *Environmental regulations:* EPA (2009) has suggested that a green community plan should incorporate the major environmental laws and regulations from federal, state, and local governments as a legitimate base for local environmental management.
- *Ecosystem concept:* It has been recognized as the most important theoretical concept for effective environmental management at the local level. The concept of ecosystem should involve its functions, processes, and integrity.
- *Rare, threatened and endangered species:* The keystone to flora and fauna protection at the local level.
- *Ecologically sensitive lands:* An eco-city and green community plan should identify lands with significant ecological value such as important vegetation, forestry, wildlife habitat, and wetlands.
- *Water resources:* This element includes three types of water resources: surface water, ground water, and imported supplies. An eco-city or green community should identify water resources being used for urban, agricultural, and environmental purposes. Additionally, the plan needs to analyze water-use situations and anticipate future trends at the community level.
- *Water quality:* An eco-city and green community plan must identify the potential problems in water quality. In general, water quality problems come from two different sources: point pollution and non-point pollution.
- *Ground-water depletion:* In many cities, water supplies are highly dependent on ground water; at the same time, urban development causes a huge negative impact for this source of water. Unlimited ground-water use in urban areas will cause a long-term decline in the aquifer water-level. Many negative consequences result from ground water depletion such as drying up of groundwater wells, reduction of surface water, deterioration of water quality, increased pumping costs, land subsidence and salt water intrusion in coastal areas.
- *Urban hydrological regimes:* These may be affected by urban land use. The impacts can be observed in rivers, streams, drainages and natural or urban aquatic resources.
- *Environmentally vulnerable lands:* This element is somewhat different from ecologically important lands. Tang's (2009) research noted that environmentally vulnerable lands may include some natural or built environment (e.g. airports; coastal zones; areas susceptible to flooding and geologic or seismic hazards and fires; areas of special biological significance; and areas of special cultural significance).
- *Wetlands:* U.S. EPA and many other studies have highlighted the value of wetlands which provide numerous benefits for the natural environment, built environment, and human health. Wetlands protect and improve urban water quality, provide fishing and recreation areas, provide wildlife habitat, store urban runoff and floodwaters, and keep communities healthy.
- *Urban vegetation and forestry:* This item relates directly to the urban human environment by providing shade, beauty, and privacy for communities and residents. Urban vegetation and forestry provide habitat for species, sources for fresh and clean air, all important components of urban ecosystems.
- *Air pollution:* One of the most important natural resources is the atmosphere. Air pollution has been widely identified as a major environmental problem in urban areas.
- *Greenhouse gas emission:* Urban areas are the major source of greenhouse gas emissions that can cause global warming. At the same time, local urban planning is an important approach to reduce greenhouse gas emission.

- *Ozone layer depletion:* This is a global environmental problem which causes a negative impact on human health. Local community policies are necessary to ameliorate this problem since the main sources come from local air conditioners, refrigerators, etc.
- *Climate change and variability*: Climate change is closely related to local human activities such as energy consumption and waste production. Thus local communities can play a critical role in climate change mitigation and adaptation.
- *Agricultural lands:* Agricultural lands can provide local food to significantly reduce the ecological footprint in food transport. Agricultural lands are a key resource in eco-city and green community planning.
- *Open space and green space:* Open space and green space have important ecological functions in groundwater recharge, wildlife habitat, flood protection, clean air, carbon offsets and climate control.
- *Historical heritage:* The historical and cultural heritage helps cities and communities maintain their identity.
- *Population change trends:* Population growth is the major factor in local environmental conflict. Population change projections can provide accurate scientific information to calculate a community's carrying capacity.
- *Noise sources and places:* Noise has a negative impact on living environment quality and human health. An eco-city and green community plan needs to identify major noise sources and sensitive areas.
- *Community risks and vulnerability:* Natural and man-made hazards put communities at risk; therefore communities need to identify major risks, vulnerable area and populations.

After an eco-city and green community plan builds a factual basis, the planning decision makers, stakeholders, and citizens can work together to develop appropriate strategies, tools, and policies to achieve sustainable community goals. The following section lists major urban planning policies for eco-city and green community planning.

- *On-site environmental assessment:* This is a site-level detailed investigation to decide whether the project matches established environmental regulations and standards.
- *Environmental threshold analysis:* This tool conducts a quantitative analysis for a project's significant environmental threshold to be used for determining potential environmental effects and ensuring compliance with specific standards.
- *Spatial analysis:* Geographical Information System (GIS) has been widely used in environmentally spatial analysis to analyze environmental impacts at the geographic scale and find solutions from an ecosystem perspective.
- *Scenario or trend analysis:* It provides alternative choices based on different scenarios. This method can be used for middle-term or long-term urban development and environmental analysis since there are always many uncertain factors influencing community development. It can be used to predict the status of a resource, ecosystem, population, transportation, or long term land development.
- *Network analysis:* This method uses a systematic evaluation to analyze indirect and cumulative environmental effects from community development. It can be used for ecosystem analysis to identify complex cause-and-effect relationships.
- *Ecological footprint analysis:* It is a recent emerging approach to analyze the relationship between consumptions of natural resources and waste disposal. It can be

used to calculate residential demands under a certain environmental carrying capacity.

- *Questionnaires, interviews, and expert panels:* One of the most effective approaches to get opinions and feedbacks from stakeholders and citizens regarding projects, programs, and plans.
- *Checklist and matrices:* This is a simple but effective approach to identify cause and effect and determine the inter-relationships between human development and environmental impacts.
- *Life-cycle analysis:* This approach focuses on the complete life cycle of a project or product enabling communities to develop appropriate policies or strategies to address environmental impacts from start to finish.
- *Compatibility appraisal:* This method determines whether a project is compatible with established environmental laws and regulations. It can also be used to test a policy's internal coherency and consistency.
- *Benefits and costs analysis:* It helps make economically feasible planning decisions.
- *Risk and vulnerability analysis:* This method analyzes the probability and consequences of community risks in regard to vulnerable geographic areas and population groups.
- *Land use restriction and permitting use:* The most common rational planning policy to help protect important environments.
- *Density requirement:* Density requirements create a dense and environmentally friendly society.
- *Buffer or setting back requirements:* This policy protects important environmental resources, softens landscape views, protects privacy, and mitigates noise.
- *Protection or special zoning:* This policy reserves important lands for research, education, or conservation. For example, a special zone can be set aside for aquifer protection or biodiversity habitat.
- *Growth boundary control:* This policy is designed to control urban sprawl by setting strict urban growth boundaries to protect open space, promote in-filled and mixed use, and save costs.
- *Disaster-resistant land use and building codes:* This policy sets a series of standards, permits, and codes for building a disaster-resistant community.
- *Transfer of development rights:* Communities can create mechanisms to permit the transfer of development rights from designated places to nearby or other places thereby preserving farmland, open spaces, and sensitive areas.
- *Conservation and mitigation banking:* This policy creates some lands to be held for future use. It can be used for wetland restoration, open space protection, and protection of water supply areas.
- *Density bonus or bonus zoning:* This policy encourages communities to create higher density development patterns.
- *Clustered development:* It can change the trend of urban sprawl and allow development on only certain lands while conserving remaining land as open space.
- *In-filled development:* This policy is designed to rehabilitate, maintain, and improve existing land rather than use new natural land and create urban sprawl. Addresses existing infrastructure and appropriate reuse and redevelopment of previously developed or underutilized lands.
- *Mixed use and compact development:* This policy has become an attractive approach to control urban growth by reducing auto dependence, reducing infrastructure costs,

preserving green space and natural resources, promoting revitalization and social equity.

- *Pedestrian/resident-friendly, bicycle-friendly, transit-oriented development:* It is an important policy for building livable, active, clean, healthy, energy-conserving communities.
- *Recycling programs:* This policy saves natural resources (e.g. wood, water, and minerals) and conserve energy. These programs can reduce greenhouse gas emission because fewer new products need to be made. They also save money for residents and the community and prevent the destruction of natural habitat and land.
- *Low-impact design:* Low impact design for impervious surfaces brings a number of ecological benefits for surface water quality, groundwater recharge, watershed and wetland ecosystems, and flood prevention.
- *Watershed-based and ecosystem-based land management:* This policy puts local environmental issues in a regional context to find collaborative solutions.
- *Leadership in Energy and Environmental Design (LEED):* This policy and standards can create more efficient, healthy, active living environments by improving air quality, saving water resources, reducing solid waste, and saving the money as a whole.
- *Public participation and stakeholder involvement:* The current planning theory of communication and collaboration encourages active public and stakeholder involvement in community decision making by creating more accountable, transparent, and active decisions and reducing potential conflicts.

The above section listed the most frequent policies and strategies that can be used to create eco-city and green community. Cities and communities may adopt more than one of the policies, tools, and strategies for urban planning and environmental management.

CONCLUSIONS

Although eco-city and green community is seen more in theory than in practice, it suggests that many traditional, unsustainable, unhealthy development patterns of the past should be corrected or avoided in the future. Our first section defined what an eco-city and green community is and reviewed the history of the concept of eco-city and green community. Our second section offers evidence of the sustainability of different urban forms. It does not endorse a certain urban form, but the discussions of each type of urban form helps to understand urban development patterns. The following table (Table 1b) provides a summary of the major models discussed in this section.

After comparing these models, it appears that the dense city model offers a more flexible structure for changing cities. The linear metabolism city model is perhaps the more ideal model which emphasizes the use and reuse of resources. The integrated macro-and micro structure of the model is useful for evaluating a city's potential for achieving a sustainable urban form. Comparing the four physical models – linear metabolism city model, compact city model, the macro-structure of alternative city models, and the dense city model -- it is apparent that each type of model has its own strengths and weaknesses. Therefore, it is important to point out that all of those models can be used under particular conditions. Each of the models can focus on different aspect of sustainable cities. In practice, a more general model, rather than a single model can be integrated with other types of models. The compact city model has been well analyzed and widely adopted in many cities. The obvious advantages have been summarized in previous research (Breheny, 1992; Jenks et al. 1996). The compact city model is an ideal urban form in many respects: conservation of the countryside; efficient use of existing lands; protection of important natural resources; less need to travel by car, thus reduced fuel emissions; support for public transport and walking and cycling; better access to services and facilities; more efficient utility and infrastructure provisions; and revitalization and regeneration of inner urban areas. However, to date, some critics have pointed out that there is little evidence to support such benefits, and the sustainability of the compact city continues to be questionable, while many urban researchers expand the idea of the compact city model which is the origin of the macro-structure of alternative city model (Frey, 1999) and the dense city model (Rogers, 1997) as well. These models offer the choice of a planning design to improve an existing city or city region. They focus on the relationship where social activity interfaces with neighborhoods and the built environment. We believe that the dense city model with more flexible structure and clearer transport principles is currently more appropriate for today's rapidly changing cities. Last, we recognize that the physical city models all represent macro-scale forms. The goal of the actual physical city models is the overall compactness or dispersal of urban fabric which suggests the concentration or decentralization of services, facilities and workplaces within the overall city form. The key elements of eco-city and green community planning provide theoretical and practical guidance for local communities.

Table 1b. A summary of major urban form models

Explorer	Physical model investigated	Conclusions
H.Girardet (1992)	Linear metabolism city model	The main finding is that consumption should reduced by implementing efficiencies and maximizing re-use of resources.
CEC (1990)	The 'Compact City' model	This model encourages a high-density, mixed-use urban form.
H. Frey (1999)	The macro-structure of alternative city models (The Core City, The Star City, the Satellites City, The Galaxy of Settlement, The Linear City, and The Regional city models)	This investigation asks how individual models compare in terms of overall dimensions and total area required and how they respond to sustainability.
R.Rogers (1997)	The 'Dense City' model	It is a compact, well-connected flexible structure which focuses on the relationship where social activity overlaps neighborhood and the built environment.

In conclusion, the implementation of the strategies of eco-city and green community will greatly benefit cities and citizens because of significant environmental, economic, and social savings.

REFERENCES

Archibugi, F. (1997). *The ecological city and the city effect: essays on the urban planning requirements for the sustainable city,* Ashgate Press.

Barton, H. (2000). *Sustainable communities: the potential for eco-neighbourhoods,* Earthscan Publication Ltd.

Berke, P. R. (2008). The evolution of green community planning, scholarship, and practice. *Journal of the American Planning Association,* 74(4): 393-407.

Brooks, M. P. (2002). *Planning theory for practitioners,* Planners Press, Chicago, IL.

Burton, E., Jenks, M. and Williams K. (2000). *Achieving sustainable urban form,* E & FN Spon.

Campbell, H. (2006). Just Planning, The art of situated ethical judgment. *Journal of Planning Education and Research,* 26:92-106.

Elkin, T., Mclaren, D., and Hillman M. (1991). *Reviving the city: towards sustainable urban development,* London: Friends of the Earth

EPA. (2009). Green communities, http://www.epa.gov/greenkit/index.htm (visited on May 20, 2009)

Fischler, R. (2000). Linking planning theory and history: The case of development control. *Journal of Planning Education and Research,* 19: 233-241.

Fischler, R. (2000). Communicative planning theory: A foucauldian assessment. *Journal of Planning Education and Research,* 19:358-368.

Fainstein, S. S. (2005). Planning theory and the city. *Journal of Planning Education and Research,* 25:121-130.

Frey, H. (1999). *Designing the City: Towards a more sustainable urban form,* E&FN Spon.

Girardet, H. (1992). *Creating sustainable cities,* Green Books Ltd.

Girardet, H. (1996). *The Gaia atlas of cities: New directions for sustainable urban living,* Gaia Books Limited.

Girardet, H.t (2004). *Cities people planet,* Wiley-Academy.

Gosling, D. and Maitland, B. (1984). *Concept of urban design,* London: Academy Edition.

Gumuchdjian, P. (1997). *Cities for a small planet,* Faber and Faber Limited.

Hall, Peter (1988). Cities for Tomorrow: An intellectual history of urban planning and design in the twentieth century, Basil Balckwell.

Hall, P. and Pfeiffer, U. (2000). Urban future 21: A Global 21th Century Cities, E&FN Spon.

Herbert T. D. and Thomas J. C. (1982). Cities in space, city as place, David Fulton Publishers Ltd

Holden, M. (2008). Social learning in planning: Seattle's sustainable development codebooks. Progress in Planning, 69: 1-40.

Howard, E. (1898). Garden cities of tomorrow, Faber and Fafer Limited.

Huxley, M., and Yiftachel, O. (2000). New paradigm or old myopia? Unsettling the communicative turn in planning theory. *Journal of Planning Education* and Research, 19:333-342.

Hurley, P. T., and Walker, P. A. (2004). Whose vision? Conspiracy theory and land-use planning in Nevada County, California. *Environment and Planning* A, 36: 1529-1547.

Huang G. and Chen Y. (2002). Ecocity: *Theory and design approach.* Science Press, China

Innes, J. E., and Booher, D. E. (1999). Census building as role playing and Bricolage: Toward a theory of collaborative planning. *Journal of the American Planning Association,* 65(1): 9-26.

Innes, J. (1996). Planning through consensus building: A new view of the comprehensive planning ideal. *Journal of the American Planning Association, 62* (4): 460-72.

Jacobs, J. (1962). *The death and life of great American cities,* London: Jonathan Cape.

Landry, C. (2000). *The creative city: a toolkit for urban innovators,* Earthscan Publication Ltd

Lauria, M., and Wagner, J. A. (2006). What can we learn from empirical studies of planning theory? A comparative case analysis of extant literature. *Journal of Planning Education and Research,* 25:364-381.

Lawrence, D. P. (2000). Planning theories and environmental impact assessment. *Environmental Impact Assessment Review,* 20: 607–625.

Longley, P. A., and Webber, R., Li, C. (2008). The UK geography of the e-society: A national classification. *Environment and Planning A,* 40: 362-382.

Lynch, K. (1960). *The image of the city,* the MIT Press.

Lynch, K. (1984). *Good city form,* the MIT Press.

McHarg, I. L. (1969). *Design with nature,* John Wiley and Sons.

Meadows, D. H. and Forrester, J. (1972). *Limits to growth: a report for the club of Rome's project on the pedicament of mankind,* London: Pan

Mumford, Lewis (1938). *The culture of cities,* Secker & Warburg

Neuman, M. (1998). Does planning need the plan? *Journal of the American Planning Association,* 64(2): 208-220.

Pugh, C. (2000). *Sustainable cities in developing countries: theory and practice at the millennium,* Earthscan Publication Ltd.

Register, R. (1987). *Ecocity berkeley: Building cities for a healthy future,* North Atlantic Books.

Richardson, T. (2005). Environmental assessment and planning theory: Four short stories about power multiple rationality and ethics. *Environmental Impact Assessment Review,* 25: 341–365.

Rogers, R. and Power, A. (2000). *Cities for a small country,* Faber and Faber Limited.

Roseland, M. (1997). *Eco-city dimensions: Healthy community, healthy planet,* New Society Publishers.

Scott Campbell and Susan S. F. (eds.): (2003). *Readings in planning theory* (Studies in urban and social change), Second Edition, Blackwell Publishing, Malden, MA.

Tang, Z., Brody, S.D., Quinn, C., Chang, L., and Wei, T. (2010). Moving from agenda to action: Evaluating local climate change action plans. *Journal of Environmental Planning and Managemen,.* 53(1):43-62

Tang, Z., Hussey, C.M., and Wei, T. (2009). Assessing local land use planning's awareness, analysis, and actions for climate change. International *Journal of Climate Change Strategies and Managemen,*1(4):368-381

Tang, Z., and Brody, S. D. (2009). Link planning theories with factors influencing local environmental plan quality. *Environment and Planning B: Planning and Design,* 36: 522 -537.

Tang, Z. (2009). How are California local jurisdictions incorporating a strategic environmental assessment in local comprehensive land use plans? *Local Environment,* 14(4): 313-328.

Tang, Z., Bright, E., and Brody, S. D. (2009). Evaluating California local land use plans environmental impact reports. *Environmental Impact Assessment Review,* 29: 96-106.

Tang, Z. (2008). Integrating the principles of strategic environmental assessment into local comprehensive land use planning. *Journal of Environmental Assessment Policy and Management,* 10(2): 143 -171.

Tang, Z. (2008). Evaluating the capacities of local jurisdictions' coastal zone land use planning in California. *Ocean and Coastal Management,* 51(7): 544-555.

Tang, Z., Lindell, M., Prater, C., and Brody, S. D. (2008). Measuring tsunami hazard planning capacity on the U.S. Pacific coast. *Natural Hazards Review,* 9(2): 91-100.

Tibbalds, F. (1992). *Making people-friendly towns: Improving the public environment in towns and cities,* Spon Press.

Thomas, R. (2002). *Sustainable urban design: an environmental approach,* London: Spon Press.

Trancik, R. (1986). *Finding lost space: Theories of urban design,* New York: Vsn Nostrand Reinhold Company.

Urban Task Force (1999). *Towards an urban renaissance,* London: E & FN Spon.

Velasquez, E. L. (1996). Agenda 21: A form of joint environmental management in Manizales, Colombia. *The earthscan reader in sustainable cities,* Edited by Satterthwaite, David.

Wackernagel, M. and Rees W. (1996). *Our ecological footprint: Reducing human impact on the earth,* New Society Publishers.

Watson, V. (2002). Do we learn from planning practice? The contribution of the practice movement to planning theory. *Journal of Planning Education and Research,* 22:178-187.

Wheeler, M. S. and Beatley T. (2004) *The sustainable urban development reader,* Routledge.

White, R. R. (2002). *Building the ecological city,* Cambridge: Woodhead

SECTION III: MAKING ECO-CITIES A REALITY: SOME KEY DIMENSIONS FOR ECO-CITY DEVELOPMENT WITH BEST PRACTICE EXAMPLES

In: Eco-City and Green Community
Editor: Zhenghong Tang

ISBN: 978-1-60876-811-0
© 2010 Nova Science Publishers, Inc.

Chapter 3

MAKING ECO-CITIES A REALITY: SOME KEY DIMENSIONS FOR ECO-CITY DEVELOPMENT WITH BEST PRACTICE EXAMPLES

Jeffrey R. Kenworthy

Curtin University Sustainability Policy Institute,
Curtin University, Perth, Western Australia

ABSTRACT

Making existing cities and new urban development more ecologically based and more livable is becoming an increasingly mainstream planning objective as cities try to fulfil their local, national and global responsibilities for greater sustainability. The pressures of global climate change and the sceptre of peaking world oil production are two powerful forces at work in cities today. But also at work are a host of other local imperatives to ensure greater ecological orientation in urban development and environmental improvements, better social development of the city and the need to maintain a good economic base for the city. This paper discusses ten critical responses to the issue of eco-city development by way of some useful examples of each response from some cities around the world. A previously developed simple conceptual model involving ten key planning and transport factors for more ecologically based urban development is used as the basis of the chapter. The ten factors cover (1) compact, mixed use urban form, (2) protection of the city's natural areas and food producing capacity, (3) priority to the development of superior public transport systems and conditions for non-motorised modes, with minimal road capacity increases, (4) environmental technologies aiming for closed loop physical systems, (5) well defined higher density, human-oriented centres, (6) a greatly enhanced public realm throughout the city, (7) sustainable urban design principles, (8) economic growth based on creativity and innovation, (9) vision-oriented rather than "predict and provide planning" and (10) decision making within a sustainability framework involving genuine public engagement. Each one of these pursuits is briefly summarised and illustrated with some examples of best practice from around the world.

INTRODUCTION

Changing urban development from its present unsustainable forms and patterns in both wealthy and poorer cities is a very challenging process. Not only do urban form, transportation systems and water, waste and energy technologies have to change, but the value systems and underlying processes of urban governance and planning need to be reformed to better reflect a commitment to sustainability.

This paper briefly summarises ten critical responses to the issue of changing the nature of urban development to a more ecological, sustainable model. These dimensions revolve around urban transport systems and their links to urban form and are therefore mostly, though not exclusively, focussed on the problems of reducing automobile dependence in cities, building more sustainable urban form and creating more livable places.

These ten dimensions are not exclusive of other critical factors in the quest for urban sustainability and some caveats, limitations and omissions, as well as a detailed description of each dimension, have been provided in Kenworthy (2006). However, these ten dimensions are central to any attempts at greater sustainability in both prosperous and less prosperous cities, especially because of the powerful city-shaping ability of transport systems.

Before providing some examples of these dimensions in action, a brief summary of each is now outlined.

THE TEN ECO-CITY DIMENSIONS

The ten critical eco-city dimensions examined in this chapter, along with a brief description of each, are as follows:

1) *Compact, mixed-use urban form that uses land efficiently and protects the natural environment, biodiversity and food producing areas.* There is extensive literature showing the positive effects of compact, higher density, mixed use urban form on reducing dependence on cars and increasing the convenience and quality of public transport and mixed land use (Holtzclaw, 1994; Kenworthy and Laube, 1999; Newman and Kenworthy, 1989, 1999; Kenworthy and Laube 2001; Naess, 1993a,b). In essence, higher density, mixed uses provide shorter travel distances that make walking and cycling more feasible and travel by transit more competitive with the car. They also provide the concentrations of people that are necessary to create the high loads needed by transit systems to offer high frequency, convenient services. Unlike road systems, which deteriorate with increasing use, transit systems typically improve as demand increases (apart from comfort issues which can be challenging in peak periods if insufficient capacity is provided).

2) *Natural environments permeate the city's spaces and embrace the city, while the city and its hinterland provide a major proportion of its food needs.* With the peaking of world oil production and the high embodied energy costs in food, it is becoming increasingly important to source food supplies more from local bioregions (Campbell and Laherrere, 1995). In the USA, "100- mile restaurants", which source all food products within a 160 km radius, are becoming popular (Beatley, 2005). Additionally, if higher densities become a reality in more cities, then there needs to be more shared green space available to inhabitants in order that people can maintain

contact with nature and have places to recreate and refresh themselves. Somewhat paradoxically, more compact urban development provides greater opportunities for "green cities", both in the sense of food growing in and around the city and more parks and natural vegetation, because not all land needs to be consumed by sprawl, roads and car parks (Beatley, 2000).

3) *Freeway and road infrastructure are de-emphasised in favour of transit, walking and cycling infrastructure, with a special emphasis on rail. Car and motorcycle use are minimised.* The need to reduce motorisation in cities and the many negative effects of excessive automobile dependence are clear (Newman and Kenworthy, 1999). Evidence about the deleterious effects of freeways in terms of increasing car use, energy consumption and automotive emissions has been available for many years (Watt, 1974). Furthermore, the futility of attempting to tackle traffic congestion and reducing fuel use and emissions through more road construction and smoother, freer-flowing traffic is well documented (Newman and Kenworthy, 1984, 1988; Goodwin, 1997). Conversely, it has been shown how significant urban rail systems are central to building better public transport systems and also improving many other important dimensions of cities, including increased levels of walking and cycling, reduced parking, fewer transport deaths and so on (Kenworthy, 2008).

4) *There is extensive use of environmental technologies for water, energy and waste management – the city's life support systems become closed loop systems.* It is very clear that the current patterns of urban resource use and waste production are unsustainable, resulting in huge urban ecological footprints (Rees, Wackernagel and Testemale, 1998). They tend to give urban systems a "parasitic" character. Cities therefore need to progressively adopt urban infrastructure systems that use renewable energy, which harvest and re-use water in a sustainable way and which feed wastes back into resource use cycles, where, as far as is possible, most life-support systems in cities become closed-loop systems. The overall aim of environmental technologies is therefore to maximise the possibility that cities can meet their needs from the natural capital of their own bioregions in a renewable way (Newman and Jennings, 2008).

5) *The central city and sub-centres within the city are human centres that emphasise non-auto access and circulation and absorb a high proportion of employment and residential growth.* Amongst the most important parts of any city are its CBD and sub-centres (Monheim, 1988; Gehl and Gemzøe, 1996; Jacobs, 1961). Central cities still remain the single biggest concentrations of jobs in most cities, despite the suburbanisation of work and the falling percentage of people employed in them (Kenworthy and Laube, 2001; Kenworthy, 2009). The high and generally increasing number of jobs and floor-space means that the central city still significantly shapes transport patterns. Public transport systems, especially rail, are focussed on central cities and congestion on radial routes is widespread. Sub-centres around cities are also crucial to making the city a more transit-oriented environment through development of a polycentric structure with "decentralised concentration" of land uses. These high concentrations of activities occur mostly around urban railway stations as in, for example, Vancouver and Stockholm (Cervero, 1998).

6) *A high quality public realm in the city, which expresses a public culture, community, equity and good governance. The public realm includes the entire transit system and all the environments associated with it.* A compelling factor that distinguishes 'good' cities from 'bad' cities is how they address the public realm (Newman, 1990). Mike Davis writes about urban communities that have abandoned their sense of responsibility concerning 'the commons', the most obvious being shared urban

spaces, streets, parks, transit systems and so on (Davis, 1990). He suggests Los Angeles has become a highly privatised, fear-driven environment, which he characterises as "The Ecology of Fear" or "Fortress LA". Though an extreme example, Los Angeles demonstrates the wider general proposition that the public realm in cities, especially the streets, is crucial in making them more livable and sustainable and indeed democratic (Barber, 1995; Appleyard, 1981; Jacobs, 1993). Putnam (2001) implies that a good public realm is also critical in the development of social capital. One of the features of much current development in Los Angeles, which is occurring around stations on its growing urban rail system (light rail, metro and commuter rail systems), is that the public realm is becoming more attractive and inclusive and less "fear-driven".

7) *The physical structure and urban design of the city, especially its public environments are highly legible, permeable, robust, varied, rich, visually appropriate and personalised for human needs.* The physical layouts and designs that make the most enduring and loved cities have long been known. A range of authors provide detailed accounts of the design of Greek, Roman, Chinese, Japanese and new world cities such as Boston and Los Angeles, showing the central importance of, for example, permeable street patterns, based on regular or deformed grids and legible streetscapes punctuated by well-placed landmarks and significant buildings [Lynch, 1960, 1981]; Kostoff, 1991; Bacon, 1974). Others have developed a suite of measurable design qualities that need to be incorporated into urban development (Bentley et al, 1985). These principles reflect centuries of wisdom in place making, which automobile cities have largely ignored, but are now rediscovering through movements such as The New Urbanism (Katz, 1994).

8) *The economic performance of the city and employment creation are maximised through innovation, creativity and the uniqueness of the local environment, culture and history, as well as the high environmental and social quality of the city's public environments.* Jane Jacobs shows that cities are the key sites and drivers of national economies and cities themselves cannot survive without a viable economic base (Jacobs, 1969, 1984). Any city aspiring to sustainability cannot ignore its economic dimension. Since Jacobs' time, globalisation has strengthened the role of cities in driving the global economy. The "creative city" approach, whereby cities attempt to find innovative and often more locally-based ways of diversifying and expanding their economies, is now an accepted means of economic progress and innovation (Landry, 2000; Florida, 2002, 2004).

9) *Planning for the future of the city is a visionary 'debate and decide' process, not a 'predict and provide', computer-driven process.* In the post-World War II period transport planning has been characterised by the use of computer models designed to predict future traffic growth and to work out how much new road infrastructure will be required to meet that projected demand (derogatorily referred to as "predict and provide"). This approach has had very negative effects on the environment of cities, first in cities of the west, and now increasingly in rapidly developing poorer cities where the approach is still being aggressively applied. This is at a time when its use is seriously waning in the West in favour of a transportation demand management (TDM) approach, which attempts to match transportation demand to existing infrastructure provision. This predict and provide technical methodology and the accompanying political and policy decision making processes have resulted in construction of extensive freeway systems, increasing the level of the car travel and energy demand and generating spiralling emissions (Mitchell and Rapkin, 1954; Kenworthy, 1990). The sustainability demands in cities are fostering a more

community-based approach of envisioning the future city and asking "what do we want our city to look like in 20 years from now? What qualities should it have compared today? How should it change?" This is a "debate and decide" approach.

10) *All decision-making is sustainability-based, integrating social, economic, environmental and cultural considerations, as well as compact, transit-oriented urban form principles. Such decision-making processes are democratic, inclusive, empowering and engendering of hope.* It is not surprising that for sustainable development to be implemented, quite radical departures from normal planning and decision-making processes in cities will be required. This is why there are many activities in cities around the world that are establishing visions of sustainable development and how these can be realised. The key defining characteristics of these efforts are their engagement with diverse 'communities' or 'stakeholders' that constitute any city today and their capacity to infuse a new sense of hope about urban futures (Newman and Jennings, 2008).

SOME EXAMPLES OF BEST PRACTICE

This section sets out some examples of best practice in the above key areas for making cities more sustainable. Rather than taking each area and treating them as separate subjects, it makes more sense to combine them where particular cities effectively demonstrate more than one dimension. This is also conceptually logical, since one would expect that the dimensions are linked. For example, the de-emphasising of road systems development and prioritising construction of higher quality transit systems, is often linked with the development of higher density sub-centres with a better quality public realm.

The following discussion thus highlights a range of examples of cities that demonstrate to a reasonably high degree, the principles outlined in the previous section. These cities are by no means the only cities implementing these principles of urban development, but space does not permit a wider elaboration.

Vancouver, British Columbia

Vancouver is a metropolitan area of some 2.2 million people. Perhaps its most distinguishing feature in a North American context is that the City of Vancouver at the core of the region (~578,000 or more than 25% of the population) has no urban freeways.

Vancouver has become to urban planners what Stockholm achieved in the 1950s when it built its Tunnelbana or modern metro system and proceeded to build whole communities of satellite towns around its stations such as Valingby and Kista. Stockholm became an official global pilgrimage site for planners to view the best in what is now called transit-oriented development or TOD, complete separation of cars from pedestrians and cyclists and other excellent urban design features to encourage walking and cycling.

Vancouver today is a fine example of a city that is implementing not only higher compact mixed use urban form (Dimension 1), but has also prioritised transit development, especially new rail infrastructure over new road infrastructure (Dimension 3). It has built up the role of its CBD area and especially transit-oriented sub-centres across its metropolitan region (Dimension 5), and in the process of implementing these other objectives, it has dramatically improved the quality of its public realm (Dimension 6) and has implemented sustainable urban design qualities throughout much of the urban area (Dimension 7).

Vancouver's achievements in these five dimensions are now discussed, although it also clearly has made contributions in the area of environmental technologies through the South East False Creek development, part of which is currently nearing completion and will act as the site of the new Winter Olympics Village for athletes in 2010 (Dimension 4).

Vancouver has also made a significant attempt to preserve precious agricultural land and natural areas around the city (Dimension 2). The more detailed achievements in Vancouver described below have been partly shaped by the Greater Vancouver's regional planning strategy (Livable Region), which effectively creates a green belt for the region and limits the amount of suburban land that can be developed. This is on top of an already topographically constrained city due to mountainous terrain and the narrow Fraser River valley.

Vancouver has also done research that demonstrates the link between compact urban form, economic performance and livability through a study of neighbourhoods across the region. The BC Sprawl Report 2004 used indicators of urban form, economic vitality and livability to compare neighborhoods across the Vancouver region (Alexander, Tomalty and Anielski, 2004). It found a statistically significant positive link between higher densities and mixed uses, positive economic features and enhanced livability, which suggest a three-way winning scenario for policies that are aimed at creating less auto-dependent living and more walkable and sociable environments (Dimension 8).

Finally, Vancouver has also used the principles of "debate and decide" (Dimension 9) and integrated sustainability decision making (Dimension 10) in much of its activities over the last 30 years or so. The scrapping of all freeway plans in the 1970s, the planning and consultation regarding development of centres around new Skytrain stations (see more detailed discussion below), and the new "eco-density" concept (http://www.vancouver-ecodensity.ca/content.

php?id=48 accessed June 21, 2009), have all involved, to various degrees, considerable community involvement and commitment to sustainability in decision making.

Dimension 1 and 5 – Mixed-use Urban Form and Transit-oriented Centres

Vancouver has perhaps one the most dynamic and lively central and inner city populations of any city in the auto-dependent world at places such as False Creek North and South, Yaletown, the city's West End, the Coal Harbor Redevelopment, as well as many other sites throughout the region (e.g the Arbutus Lands and Fraserlands developments). The absence of high-speed road travel has meant that premium locations near to the heart of most amenities and speedy transit have become the most popular places to live in order to maintain accessibility and acceptable daily travel times. The Vancouver region's average road traffic speed in 2006 was only 38.6 km/h, whereas metro areas in the USA and Australia average between 43 km/h and 52 km/h (Kenworthy, 2009).

By way of data to substantiate the implementation of this dimension, reference to my updated global cities database provides the necessary evidence. Vancouver had trends typical of North American cities between 1961 and 1981. Urban density declined from 24.9 per ha in 1961 to 21.6 per ha in 1971 to 18.4 per ha in 1981. Then at the time that its strong reurbanisation policies began to cut in (discussed later), assisted significantly by the absence of high speed private transportation options, it started to increase in density. In 1991 it rose back to 20.8 persons per ha, then in 1996 was 21. 6 per ha and in 2006 it had exceeded its 1961 density and was registering 25.6 persons per ha, a 19% increase in 10 years (Kenworthy and Laube, 1999; Kenworthy, 2009). Land use change of this magnitude, that is focussed mainly in central and inner areas, but which generates a significant increase in density across an entire region of 2 million people, is hard to achieve, especially since some parts of the region are continuing to grow with lower suburban densities.

A major success factor of transit development in Vancouver over the last twenty-five years has been the strong efforts to integrate high density residential and mixed use development in significant nodes around selected stations on Skytrain, redevelopment of highly favoured waterfront areas such as False Creek and Coal Harbor, and even in some cases the development of strong town centres around bus-only nodes such as Port Moody. From before its inception, Skytrain's development has gone hand-in-hand with planned high density TOD from which it draws a lot of its patronage.

It has been public policy since the mid-1970s to try to concentrate as much development in transit-rich locations. Indeed public consultations with communities affected by Skytrain-linked redevelopments occurred as early as 1978, eight years ahead of the opening of the first Skytrain segment in 1986 for the Expo that occurred on the land now known as False Creek North or Yaletown. This high density has occurred at stations in Vancouver as well as neighbouring Burnaby and further down the line in New Westminster at stations such as Joyce-Collingwood, Metrotown, Edmonds and New Westminster. Such TODs are gradually reshaping the Vancouver region into a genuine polycentric "transit metropolis" (Cervero, 1998).

Park-and-ride around stations in the City of Vancouver, City of Burnaby and City of New Westminster have been expressly excluded in favour of high density uses clustered close to the station entrances. South of the Fraser River in the Surrey suburbs, park-and-ride surrounds some stations such as Surrey Central, with development set back from the station. The resulting urban design outcomes are very poor compared to those north of the Fraser.

The larger nodes on Skytrain have mixed commercial, office, residential, retail and markets within a short walk of the station. The new housing consists of quality high rise towers, 3 to 4 storey condominium style developments and townhouses. Some of the housing consists of individual housing cooperatives, which have provided more affordable housing options. The TOD at New Westminster is set along an attractive landscaped boardwalk on the Fraser River that includes playgrounds for children and extensive gardens, trees and grassed areas. The family units have inner courtyards in which families and friends congregate. The farmers market where residents do their shopping is communally-orientated with open eating areas and a more relaxed, less structured, less sterile atmosphere than a supermarket.

Some significant evidence for the preference of Vancouverites for such well-located, short distance, non-auto travel option sites comes from the Canadian Censuses of 1986 to 2006. Over this 20 year period the population of the City of Vancouver, the core of the whole region, grew from around 430,000 people to 578,041 people, an increase of over one-third and this was in the context of falling household occupancy (e.g. 2.2 persons per occupied dwelling in 1996 to 2.1 in 2001).1

As well as the obvious nodes that have sprung up within existing urbanised areas around the Skytrain in the 25 years since its opening, the re-urbanisation trend leading to the significant population increases just described, is characterised by an enormous amount of other new development along the major diesel and trolley bus lines in the city where a lot of mixed use shopping and business activities already exist (e.g. the previously mentioned Arbutus Lands development). This development consists of medium to high density housing, including shop-top housing, with special attention to the needs of families wishing to escape the car-dependent suburbs.

The jewel in the crown in this redevelopment is the staggering intensity and quality of development at False Creek located at the foot of the downtown area and serviced by frequent trolley and diesel bus services and some Skytrain stations at points near its periphery.

The Provincial Electoral District of False Creek had a population of nearly 44,000 in 2006 (http://en.wikipedia.org/wiki/Vancouver-False_Creek accessed June 21, 2009). Development at both False Creek South and North (Yaletown), as well as South East False Creek, the 2010 Winter Olympic Village site being developed at even higher densities and as an "ecological model", provide excellent examples of how to build high density transit-oriented urban villages in central locations. These areas have extensive and beautifully designed open spaces, together with adjacent mixed land uses such as markets, hotels, cultural activities, shops and restaurants (e.g. at Granville Island). There is an enormous variety in housing forms and styles in these areas, including townhouses, terraced units, medium rise and high rise apartments, with many of the earlier residential developments being cooperative housing ventures. The extensive public spaces and children's play areas are traffic free, the only direct road access on the south side of False Creek being essentially from a two-lane road at the rear of the development, with parking mostly under the buildings.

Dimension 6 and 7– The Public Realm and Sustainable Urban Design

Vancouver is not a city that has major pedestrian areas or extensive traffic calming of neighbourhoods, as in many European cities. However, the City of Vancouver has become a highly livable place characterised by an exceptional amount of human activity along lively and interesting streets and in its public spaces. For example, one of the most interesting and livable public environments is Robson Street, the long avenue that connects the downtown with Stanley Park through the West End. The sidewalks are packed with pedestrians, notwithstanding the often bumper-to-bumper traffic and high frequency trolley bus services that operate along the street. The street and its land uses are simply designed for people.

Punter (2003) describes this strong human dimension and Vancouver's detailed attention to urban design of the public realm as a hallmark of Vancouver's success as one of the world's most livable cities. This is in stark contrast to cities in the US, which all too frequently have very hostile street environments due to automobile-orientated development, particularly large freeways and interchanges.

Vancouver's transportation plans were, nevertheless, similar to US cities in the 1960s. However, through a community-led opposition process, the exclusion of all freeway development from within the City of Vancouver's borders has meant that this part of the region in particular has developed much more around transit, with a high level of walking and cycling for other trips. If freeways had been built, not only would the land that presently houses these developments have been alienated and occupied with clover-leaf freeway junctions, the quality of life around them would have also been reduced due to fumes, noise and severance.

For example, the West End of Vancouver is the second highest density residential area outside Manhattan and enjoys thriving and diverse activities along its main roads, while the grid-based, tree-lined residential streets that run across these major streets have numerous pocket parks created from selective street closures between blocks in the fine grained street grid. The area also has the extensive Stanley Park and foreshore at its doorstep.

The whole of False Creek North and South and even beyond this area is knitted together with first class, wide pedestrian and bicycle-only facilities, and it is this environment at ground level, below the often towering residential complexes, that gives people the option of sustainable transportation, as well as conviviality and convenience. The areas are exemplary in their public realm and commitment to high quality urban design.

Along the pedestrian spines there are local shops, community facilities, child-minding centres, professional suites for dentists and doctors, meeting areas, community playgrounds and sports areas, all integrated within walking or cycling distance of most residences. For a central city location, False Creek provides an exceptionally quiet oasis and yet dynamic and varied urban location for residents and the many visitors who use the area for social and recreational purposes.

Developments such as this and others such as Coal Harbor, and the ongoing evolution of Vancouver's West End, have helped to minimise Vancouver' growth in car use in inner areas by increasing transit use and making the use of non-motorised modes more feasible and attractive. The important point here is that Vancouver is limiting outward sprawl and gradually reshaping itself into a more transit-oriented region. The attraction of this way of life in Vancouver is underpinned by the quality of the public realm, which is highly livable and conducive to social interaction and recreational activities.

Apart from these waterfront areas, the Skytrain stations and high density residential precincts in the inner city of Vancouver, other focal points for high density development have been created in areas such as The Fraserlands Development connected by feeder buses to Skytrain stations. Still other very attractive and significant high density, mixed use developments have occurred in areas only serviced by buses, such as Port Moody, though there are plans to build the Evergreen Line of the Skytrain to this area (or light rail). Within these centres, pedestrians and cyclists are given attractive and comparatively safe conditions and there exists a civic life in the city spaces that is atypical for a majority of North American cities where „big box" retail centres and office parks tend to be the norm.

Dimension 3 - Transit Infrastructure, not Freeways

Vancouver's transformation from a typical auto city with a relatively short urban history compared to its east coast North American counterparts, really commenced in the early 1970s with the successful community-led fight to rid the city of all planned freeway construction within the City of Vancouver boundaries. This fight involved a then shopfront lawyer named Michael Harcourt who helped the Chinatown community to remove the threat of a freeway. He later became a Vancouver city councillor, mayor of Vancouver and finally Premier of British Columbia, a political career built significantly on fighting freeways and campaigning for more livable neighbourhoods in their place.

Transit in Vancouver has achieved a lot over the last 25 years or so. In 1981 transit use was 111 annual trips per person, which had declined to 95 per person by 1991. Then in 1996 it had risen to 118 and now in 2006 it stands at 134 trips per person per annum. This is just short of its 1961 figure of 138 trips per person, when car ownership in Vancouver was a mere 285 cars per 1000 people. Now, with almost the same transit use as in 1961, the region has 506 cars per 1000 people. Clearly, even in the context of high car ownership, transit is becoming more attractive and popular with Vancouverites due to a combination of better, speedier and more diversified services, more attractive ticket offers especially to students, but also because many more people are now living within walking distance to transit stops and feeder services to speedier rail, and bus services have improved greatly. Perhaps as a consequence, Vancouver's car ownership was even down by 3% in 2006 from its 1996 figure of 520 per 1000 (Kenworthy and Laube, 2001; Kenworthy, 2009).

Compared to most American cities Vancouver enjoys quite high levels of transit use (134 trips per person in 2006), or exactly double that of the average for 10 large US cities of 67 trips per person in 2005. The New York Tri-State metropolitan region is by far the most transit-oriented US urban area and has 168 trips per person, so that Vancouver is not so far behind this top performing US city. On the other hand, it does have a considerable way to go to catch up with some of its Canadian peers, for example the Montreal region with 206 trips per person, or even Toronto with 154. However, between 1996 and 2006 both these cities were either stagnant or slightly declining in transit use (Kenworthy, 2009), while Vancover's use of transit is still growing (13% increase from 1996 to 2006) and could overhaul these other traditional Canadian transit metropolises if such trends continue.

Vancouver's transit system consists of a comprehensive network of both diesel and trolley buses, specialised bus services for people with disabilities, an advanced, elevated and driverless LRT system called Skytrain operating at about 2 minute intervals in the peak and 6 minutes in the off-peak, a commuter rail line called the West Coast Express, servicing distant suburban communities, and a ferry service called the Sea Bus. As well, there is a new, fully automated 18.5 km partially underground, elevated and at-grade rail line from the city to the airport and a branch into the Richmond suburbs, which opened August 2009. It will operate on headways of 6 minutes, have expected ridership of 100,000 per day by 2010 and a travel time of 24 minutes (The Canada Line: www.canadaline.ca). If we assume a more modest patronage of 80,000 per day over the whole 365 days and a population of 2.3 million, this line alone will add about 13 transit trips per capita in Vancouver (an increase of 10%), excluding increased associated new boardings on access transit lines.

Within the City of Vancouver, which is built on a fine-grained traditional street grid, the bus system is relatively frequent with buses operating north-south and east-west, providing good radial and cross-city travel opportunities, with speedier services called the B-Line in some areas. Overall, however, bus average speeds are mostly below 20 km/h due to frequent stops, many intersections, (because of the fine grained grid along which they operate), moderate traffic congestion and passenger loads that are often horrendously high, affording poor passenger comfort. The buses do, however, interconnect well with each other in many locations and transfers to Skytrain, the West Coast Express, the Sea Bus and the soon-to-be Canada Line are also well-catered for, meaning that mobility in all directions across the whole region on transit is feasible without necessarily passing through the central area, and transit trips are often competitive in speed terms. The lack of freeways means that although the bus system is often slow, travel speeds can still be competitive with the car, especially where a rail segment or B-line bus is involved.

Other cities that have demonstrated similar achievements to Vancouver in these five dimensions are Stockholm in Sweden, which is one of the world's great "transit metropolises" (Cervero, 1995, 1998). Freiburg im Breisgau in southern Germany has also had a strong emphasis on higher density mixed use development, strong priority to transit, walking and cycling, excellent development of centres based around transit (more linear than nodal centres because of the frequent stopping light rail system in use there) and superb attention to the public realm and urban design in the form of an extensive pedestrianised central area and widely traffic calmed inner city. It was referred to as early as 1989 as the Green Planners Dream (TEST, 1989). Also in the German-speaking world, it is difficult not to mention cities such as Zurich, Bern and Munich for their superb transit systems and very attractive public realms, urban design, development of high quality centres and a commitment to raising densities and mixed land uses for more effective and convenient transit, walking and cycling.

Freiburg im Breisgau, Germany

Freiburg is not only noted for achievements in the above five areas, but also in its attention to ensuring minimal loss of agricultural land and natural areas to urban development (Dimension 2) and in particular its achievements in the area of the development and deployment of environmental technologies (Dimension 4). This latter factor has also been important for Freiburg's economy, which now generates significant money from its environmental technology reputation and demonstration projects.

Dimension 2 – Food Growing and Natural Areas

Freiburg im Breisgau is a small university city of 219,345 people (June 30, 2008: www.freiburg.de/servlet/PB/menu/1156563/index.html - accessed June 9, 2009) nestled in the Black Forest area of southern Germany. The total land area of the city of Freiburg is 15,306 ha. Of this area only 4,830 ha or 32% is required for urban development including all transportation functions. Some 6,400 ha or 42% is devoted to forests, with the remaining land area of 4,076 ha or 27% devoted to other green areas in the form of agricultural lands, recreational areas, nine water protection areas and other undeveloped land (Salomon, 2009; http://www.freiburg.de/servlet/PB/menu/1156556/index.html#Flaeche, accessed August 5, 2009). This means that more than two-thirds of Freiburg's land area is devoted to green uses. As a typical German city Freiburg has an urban density of 45 persons per ha (the average for the German sample in my database is 48 per ha), which is approximately three times the density of typical auto cities in the USA and Australia, but not quite double the density of cities in Canada. It is only possible to achieve this kind of land use result where there is a commitment to stopping the spread of new urbanisation into greenfields sites and this is exactly what Freiburg does. It identifies growth areas, usually areas that are up for redevelopment, master plans those areas for higher density urban residential and mixed use development, and then connects them very well to the rest of the city with high quality public transport, mainly using LRT.

It is impossible to achieve this level of land devoted to green uses in metropolitan areas where sprawl and the automobile dominate development. As a result of Freiburg's urban development strategy, it has an extraordinarily low level of car use for meeting daily travel requirements. For *all* daily trips, in 1991 Freiburg had achieved a situation where walking and cycling together accounted for 40% of daily trips and public transport 18%, leaving only a 42% share for cars (Bratzel, 1999). By 1999 in Freiburg, 50% of all daily trips were walking and cycling, 18% transit and 32% were by car. The goal for 2020 is 51% walking and cycling, 20% transit and 29% car (Salomon, 2009).

Dimension 4 – Environmental Technologies

Freiburg has created a reputation worldwide as an environmental technology global "hotspot". In 1975 the State government of Baden-Württenberg's decided to build a nuclear reactor. Opposition to the move was very intense and successful and spawned a civil society movement with heavy university involvement to ensure that Freiburg could then meets its future energy needs in a sustainable way. The 1980s was the period of the energy supply concept for Freiburg. That meant renewable energies and it meant doing everything possible to curb demand for energy, including focussing on transit, walking and cycling and creating a built form and public realm that favoured these modes and minimised the need to travel (Peirce, 2009). In the 1990s Freiburg further responded to sustainability by basing its future development on a climate protection concept and from 2007 on a climate protection action plan that aims for 40% of current CO_2 by 2030 through a focus on sustainable transportation and building and construction standards (Salomon, 2009). Freiburg has had low energy construction since 1992, a subsidy programme to encourage the deployment of energy saving construction since 2002, and new stricter building standards since 2008, with formalised energy saving plans in all public buildings.

Freiburg's development for decades now has been strongly based on citizen action and participation. Citizens are shareholders in solar and wind power stations. There is direct participation in the spatial development plan and the municipal budget. Citizens act as technical experts on committees and there is much citizen-led environmental education and many campaigns. This citizen participation and commitment and the networks of stakeholders have helped create a vision of integral sustainable development, which enjoys a consensus across all political parties (Salomon, 2009).

Freiburg has the following achievements in this field:

- Freiburg has cogeneration covering 50% of electricity demand.
- There is district heating in new city quarters such as Vauban. Vauban uses a biomass burning plant to dispose of some of its organic waste.
- It has 90 small scale Combined Heat and Power (CHP) cogeneration plants.
- The city has five 1.8 Megawatt (MW) windmills.
- There are several small scale hydro power turbines.
- The city has more than 10 MW of installed photovoltaic power with 10 million kWh output per annum and 15,000 square metres of solar thermal infrastructure.
- The region is selling itself as a Solar City with strong research, industry and a network of stakeholders called SolarRegion Freiburg.

Freiburg also has a multi-layered waste management strategy consisting of a series of different types and colours of bins for waste separation at source (much of this is also common throughout German cities). There are regular waste bins, bio bins for bio-degradeable wastes, green paper bins for paper recycling, and yellow sacks for recyclable plastics. It has 380 stations for glass recycling (green, white and brown glass in separate containers). There is separate collection of hazardous wastes not permitted in any bins and places for the collection of electronic waste such as old computers and mobile phones. Finally it has reuse in the form of a produce exchange system.

These principles and more, such as sustainable water management and sustainable transport, are practiced in new extensions to the city such as Rieselfeld and Vauban.

The district of Freiburg-Rieselfeld was developed out of a need to provide for a very high demand for new housing in the late 1980s and early 1990s and accommodates 11,000 people. Rieselfeld was only possible through as extension of Freiburg's excellent LRT system, along which there are several stops serving the new district. The LRT runs on a grass track bed through the new district, along which is located a linear neighbourhood center with a rich mix of shops, food stores, restaurants, professional suites and other uses, and above those are several floors of housing. The whole of Rieselfeld is accessible by foot to the LRT stops and both the main street and the residential streets connecting to it are very bicycle and walking friendly and indeed at many times of the day there are a lot of children riding bikes, walking and playing in the general street environment. Along this main LRT street are civic functions such as a library, churches and a large square where in the summer, children enjoy computerised water fountains. Rieselfeld is an excellent example of TOD linked in a linear rather than nodal form to new urban development.

Sustainable water management involves the rainwater falling on the site being channelled through green swales that also act as a green open space network through the development. There are also some "holding ponds" for water that are landscape elements amongst the housing and young children can be seen pretending to fish in them. As well, many of the dwellings can be seen sporting photovoltaic arrays and solar hot water systems.

Vauban is a redevelopment area, on a site near to the city, including an old French military barracks and linked to the rest of the city by an extension of the LRT system, again running along a green track bed. Some of the old buildings have been retained and recycled into a kindergarten and other civic uses. Vauban is a dense, mixed-use new neighbourhood of 5,000 people but with the added feature that it is strongly focussed on environmental technologies, especially for renewable energy. It has passive and plus energy houses, meaning that these latter dwellings generate net energy, which is fed back into the grid and as already indicated, it has its own power plant burning waste organic material, mainly wood waste.

It is a "car free" neighbourhood, meaning that if one wants to have a car one has to store it in a solar parking structure on the fringe of the neighbourhood, which generates enough power for the garage's needs but also feeds excess power into the grid. Vauban is strongly oriented to transit, walking and cycling and one of its most evident and endearing features is its family-friendly public realm. Throughout the development men and women can be seen pushing children in prams and children can be seen independently walking and riding bikes around the area, simply because the street environments are comparatively safe with 30 km/h (19 m.p.h) residential zones. There are also many attractive parks, which are intensively used by parents with children. Overall, Vauban is probably one of the most attractive sustainable neighbourhoods in the world, successfully blending, high density housing, mixed uses, green spaces, transit and walking and facilities and the use of environmental technologies into a rich and highly livable, socially gregarious and safe public environment.

Portland, Oregon; Boulder, Colorado and Perth, Western Australia – debate and decide, not predict and provide and integrated sustainability decision making (Dimensions 9 and 10)

Portland, Oregon

Portland is perhaps the most successful and well-known example in the USA of a city that has reshaped itself under a strong vision extending as far back as the 1970s. At this time it established an urban growth boundary to limit sprawl, decided to build a light rail system call Metropolitan Area Express (MAX), which opened its first line in 1986 and scrapped a freeway that would have destroyed 3000 homes (Newman and Kenworthy, 1999). It also introduced in 2001 the first phase of a separate tram or streetcar system, which operates in the inner city area. LRT stations and areas around streetcar stops are now the major focus for all new growth in the Portland region, with numerous attractive, compact, mixed use centres developing along an extended LRT system and new tram system in the inner city, particularly in the Pearl District. Parks and green spaces have been created and property values have risen on the back of strong population growth and company location decisions. Portland had a visioning process called Region 2040, a broad-based community representation process involving 44 stakeholders developing a vision and strategic goals for the region. At the heart of Portland's growing success over many years has been strong community engagement and empowerment through groups such as 1000 Friends of Oregon who have fought for a sustainability-based vision for their region, focussed on reducing automobile dependence and radically improving transit options.

The roots of Portland's performance over the last 25 years or more in land use and transportation development dates back to the 1970s when Governor Tom McCall spearheaded a statewide growth management strategy, in particular the establishment and maintenance of urban growth boundaries (UGBs) in Oregon. Portland established such a boundary inside which all urban growth had to occur. This boundary has been linked to transportation through the „Oregon Transportation Planning Rule", from 1991 which applies a growth rule to limit increases in Vehicle Miles of Travel (VMT). The status report in Spring 2009 from Tri-Met indicates that between 1996 and 2006 daily VMT increased only 19% in the face of a 27% increase in population, while transit patronage rose 46% (Tri-Met, 2009). Portland in 2007 ranked 7th nationally in per capita transit ridership, behind only Chicago, Boston, Honolulu, Washington DC, San Francisco and New York.

The beginnings of the comparatively strong transit success of Portland in a US context can also be traced to the land use-transportation integration evident in Portland's 1973 Downtown Plan. A Transit Mall, which opened in 1978 was envisioned as the centrepiece of downtown revitalization and marked the beginnings of a trend to leverage broader community building objectives through transit investment. Other achievements began to punctuate where Portland was going as a city: the conversion of a downtown parking lot to a park, the creation of Pioneer Square out of a parking lot, now a major community meeting point in downtown between the one-way pair of streets along which MAX operates, the tearing up of a freeway spur along the Willamette River in downtown and conversion of it to Tom McCall Park, the site of Portland's annual Rose Festival. The River Place urban village redevelopment adjacent to the new park, and itself built on a commuter parking lot, sprung up along the river with a hotel, shops at ground level and several floors of apartments above.

Further important steps in Portland's efforts to reinvent itself as a more sustainable and livable city came with successful civil society opposition to the Western Bypass loop of the I-5 freeway through rural lands just outside the growth boundary. Together with the growth management advocacy organization 1000 Friends of Oregon a study was undertaken jointly with the USEPA to develop a new approach to the problem, which culminated in a new planning model called LUTRAQ (Land Use, Transportation and Air Quality). The freeway was scrapped and now transit-oriented development is evident on the Westside light rail line opened in 1998 (at the time of opening 7,000 transit-supportive residential units were already under construction in station precincts (Arrington, 2009). Today Portland has 44 miles or 71 km of LRT and 4 miles (6.4 km) of streetcar. Along this network over $US9 billion of development has occurred using transit-friendly land use planning (Arrington, 2009). Portland has arguably the most aggressive approach of any US city to TOD. Its 2040 Growth Management Strategy of "build up, not out" is built around transit.

Perth, Western Australia

Perth, Western Australia engaged in a community visioning process in 2003 called "Dialogue With The City", which evolved out of a broader State Sustainability Strategy involving 42 areas of government, together with business and civil society. The human settlements part of this strategy emphasises innovative and efficient use of resources, less waste output, enhanced equity and livability and a greater sense of place in local communities.

Faced with a huge increase in urban sprawl and car dependence, the State government decided to involve the community on an unprecedented scale to develop a future vision for Perth for 2030. The process involved a community survey of over 1700 households and one-day forum involving 1000 participants. A critical part of the forum was a game that each group of 10 people played to plan for the expected increase in population. Each decision taken had a flow on effect, which was either positive or negative. People were thus forced to confront the dilemmas of urban planning, trading-off personal lifestyle preferences with systems effects, such as loss of bushland, traffic congestion and other implications. All results were recorded and a final report can be found at http://www.dpi.wa.gov.au/dialogue/finalproc.pdf. This process has resulted in a new plan called "The Network City," which calls for around 60% of new dwelling construction within existing built up areas to reduce car dependence and sprawl. The process forced participants to consider the social, economic and environmental considerations wrapped up in all urban planning, i.e. to adopt sustainability-based thinking.

Perth has also been aggressive in the construction of new rail lines (about 110 km since 1993) which has involved over the 16 years a great deal of citizen engagment and very often heated debate. It has also developed a Livable Neighbourhoods design code for new urban development which tries to implement walkable communities by changing the regulatory basis of suburban subdivisions to reflect different priorities such as connectivity of road systems, a sense of place and higher densities and mixed uses in new town centres based around improved transit.

Boulder, Colorado

Boulder is a small university town with a 2005 population of 83,432 people situated in the larger Boulder County area of 271,934 people [http://factfinder.census.gov/servlet/ ACSSAFFFacts?_event=Search&geo_id=&_geoContext=&_street=&_county=Boulder&_cit yTown=Boulder&_state=04000US08&_zip=&_lang=en&_sse=on&pctxt=fph&pgsl=010 (accessed May 27, 2009)]. Boulder County is part of the larger and very highly automobile dependent Denver metropolitan region of about 2.5 million people, though Boulder separates itself from Denver's urban sprawl by a strong green belt. The city has a very progressive history in sustainable settlement terms, being the first US community, in 1959, to introduce such a green belt to prevent both its own urban sprawl and the urban encroachment around it. It did this through an innovative community organization called Plan Boulder, which still maintains an active involvement in Boulder's development (http://www.planboulder county.org/ [accessed May 27, 2009)]. Boulder has set itself apart from nearly all American communities of its size in a number of ways, but perhaps the most interesting is the way it has developed its innovative transit system.

As a whole, Boulder maintained a traditional "predict and provide", supply-side road building approach to future transportation development through most of the post-war period. However, financial, political and physical realities intervened to make this approach unsustainable. In 1996 the Transportation Master Plan set a transportation demand management (TDM) goal to hold traffic levels to 1994 levels and to reduce single-occupancy vehicle (SOV) mode share to 25% (Havlick, 2005). To do this the city focussed on travel choices, rather than being locked into compulsory car use for a majority of trips. This involved improving transit services, creating demand for transit trips, enhancing the bicycle and pedestrian system, marketing and providing good information about the new choices, changing land use and urban design approaches, tackling parking pricing and establishing some dynamic relationships between the city and the University of Colorado (CU - a "town-gown" partnership), all in consultation with the active Boulder community. It also involved stopping some large-scale road expansion in neighbouring counties, mainly through public purchase of properties and development rights to prevent development that did not fit with Boulder's growth management and TDM strategies.

Probably the most community-based innovation in Boulder has been the Community Transit Network (CTN), a network of six differently branded types of bus routes that are part of the GO Boulder network whose goal is to shift 19% of commuters from their cars onto other modes. In 1990 CTN transit ridership was reportedly 5,000 per day and by 2002 had risen to 26,000 per day or a 420 % increase (Bruun, 2004). The routes are referred to as the Hop, Skip, Jump, Bound, Stampede and Dash and buses are accordingly branded and are of different sizes. All carry bicycles at the front. The Hop services are the shortest routes and the service distances get progressively greater towards the Dash.

The Community Transit Network is, as the name implies, the product of a community consultation process. Boulder undertook its highly successful transit innovations (defining certain major corridors as high-frequency all-day, all-evening, and weekend services) only after a year of extensive public involvement. A broad-based citizens group of some 50 community leaders, working with several City of Boulder and transit agency staff, devoted a great deal of time and energy to this effort, with larger public meetings being held as well. Significant changes to transit need to be undertaken with the enlistment of public support to ensure usage and ownership of transit systems, which was Boulder's approach here. They formed a new unit called GO Boulder as a way of circumventing the Public Works Department, which was operating on traditional principles. They also established the longstanding City of Boulder Transportation Committee, which ensured citizen interest in the issues (e.g. the 2009 Mayor, Will Toor, started his civic involvement with this committee). But it finally took approximately 10 years to establish the CTN, and each route involved about 1 year of citizen-involved planner per route.

The CTN is now a well-supported community-based design using buses that are family-friendly and bus drivers are employed as community ambassadors. Strong transit use was developed through innovative Ecopass programs which give purchasers unlimited transit access at cost-effective rates, marketing and education, seamless interfaces between bus, bike and pedestrian facilities, good connections to regional services and transit-supportive land use and urban design (Bruun, 2004).

CONCLUSIONS

There are many principles that need to be used to deliver truly ecologically based or sustainable urban and transportation planning. The ten dimensions discussed in this chapter are by no means exhaustive but are certainly central to efforts towards greater urban sustainability. Each year there are more and more cities that sign up to innovative programmes to implement and practice such principles. This chapter has provided some insights into Vancouver, Portland, Freiburg im Breisgau, Boulder and Perth who have practiced various combinations of these principles with reasonable degrees of success.

Author: *Jeffrey Kenworthy* is Professor in Sustainable Cities in the Curtin University Sustainability Policy Institute (CUSP) at Curtin University in Perth. He is best known for his international comparisons of cities around the theme of automobile dependence and developing more sustainable and livable cities. He has published extensively in the transport and planning fields for 26 years and is co-author with Peter Newman of *Sustainability and Cities: Overcoming Automobile Dependence* (1999) and *The Millennium Cities Database for Sustainable Transport* (2001) with Felix Laube.

REFERENCES

Alexander, D., Tomalty, R., and Anielski, M. (2004). *BC Sprawl Report: Economic vitality and livable communities*, Smart Growth BC, Vancouver.

Appleyard, D. (1981). *Livable Streets*, University of California Press, Berkeley, 382 pages.

Arrington, G.B. (2009). 'Portland's TOD evolution: From planning to lifestyle', In Curtis, C, Renne, J.L and Bertolini, L. *Transit Oriented Development: Making It Happen*. Ashgate, Surrey.

Bacon, E.N. (1974). *Design of Cities*, Penguin Books, New York, 336 pages.

Barber, J. (1995). Mending Our Lovely Metros. *The Globe and Mail,* Focus Section D, September 9, 1995.

Beatley, T. (2000). *Green Urbanism: Learning from European Cities*, Island Press, Washington DC, 491 pages.

Beatley, T. (2005). *Native to nowhere: Sustaining home and community in a global age,* Island Press, Washington DC, (408 pages).

Bentley, I., Alcock, A., Murrain, P., McGlynn, S. and Smith, G. (1985). *Responsive Environments: A Manual for Designers*, Architectural Press, Oxford, 152 pages.

Bratzel, S. (1999). Conditions of success in sustainable urban transport policy – policy change in, relatively successful "European cities". *Transport Reviews*, vol 19, no 2, pp177-190.

Bruun, E. (2004). *Community Oriented Transit Best Practices*, Working Paper 1, Independent Assessment Study of District 2 Transit Services, Alameda-Contra Costa Transit District, Transit Resource Center, Florida.

Campbell, C.J., and Laherrere, J.H. (1995). *The World's Oil Supply 1930-2050*, Petroconsultants, Geneva.

Cervero, R. (1995). Sustainable new towns: Stockholm's rail served satellites. *Cities*, 12 (1), 41-51.

Cervero, R. (1998). *The Transit Metropolis: A Global Inquiry*, Island Press, Washington DC.

Davis, M. (1990). *City of Quartz: Excavating the Future in Los Angeles*, Vintage, London, 462 pages.

Florida, R. (2002). *The Rise of The Creative Class: And How It's Transforming Work, Leisure, Community and Everyday Life*, Perseus Books Group, New York, 416 pages.

Florida, R. (2004). *Cities and The Creative Class*, Routledge, New York. 208 pages.

Gehl, J. and Gemzøe, L. (1996). *Public Spaces, Public Life*, The Danish Architectural Press and the Royal Danish Academy of Fine Arts School of Architecture Publishers, Copenhagen, 96 pages.

Goodwin, P. B. (1997). Solving congestion (when we must not build roads, increase spending, lose votes, damage the economy or harm the environment, and will never find equilibrium)", Inaugural Lecture for the Professorship of Transport Policy, University College, London, 23 October.

Havlick, S. (2005). *TDM in Boulder: A town-gown partnership*, The University of Colorado, Boulder. Powerpoint presentation.

Holtzclaw, J. (1994). "Using residential patterns and transit to decrease auto dependence and costs", Report to Natural Resources Defense Council, Washington DC.

Jacobs, A. B. (1993). *Great Streets*, The MIT Press, Massachusetts, 344 pages.

Jacobs, J. (1961). *The Death and Life of Great American Cities*, Vintage Press, New York, 474 pages.

Jacobs, J. (1969). *The Economy of Cities*, Random House, New York, 288 pages.

Jacobs, J. (1984). *Cities and The Wealth of Nations*, Penguin, Harmondsworth, 257 pages.

Katz, P. (1994). *The New Urbanism: Toward an Architecture of Community*, McGraw Hill, New York, 288 pages.

Kenworthy, J.R. (1990). "Don't shoot me I'm only the transport planner (apologies to Elton John)", in Newman, P., J. Kenworthy, and T. Lyons, (editors). *Transport Energy Conservation Policies for Australian Cities: Strategies for Reducing Automobile Dependence*, Murdoch University, Perth, Chapter 16.

Kenworthy, J. (2006). The Eco-City: Ten Key Transport and Planning Dimensions for Sustainable City Development. *Environment and Urbanization* Special Issue, 67-85, April.

Kenworthy, J. (2008). An international review of the significance of rail in developing more sustainable urban transport systems in higher income cities. *World Transport Policy and Practice*, 14 (2), 21-37.

Kenworthy, J. (2009). Unpublished update of the Millennium Cities Database for Sustainable Transport with data for 2005. See Kenworthy and Laube (2001) below.

Kenworthy, J. and Laube, F. (1999). (with Newman, P., Barter, P., Raad, T., Poboon, C., and Guia, B). *An International Sourcebook of Automobile Dependence in Cities, 1960-1990*, University Press of Colorado, Niwot, Colorado, USA, 704 pages.

Kenworthy J, and Laube F. (2001). *The Millennium Cities Database for Sustainable Transport*, International Union of Public Transport (UITP), Brussels and Institute for Sustainability and Technology Policy (ISTP), Perth: CD ROM database.

Kostoff, S. (1991). *The City Shaped: Urban Patterns and Meanings Through History*, Thames and Hudson, London, 352 pages.

Landry, C. (2000). *The Creative City: A Toolkit for Urban Innovators*, Earthscan Publications, London, 300 pages.

Lynch, K. (1960). *The Image of The City*, The MIT Press, Cambridge, Massachusetts, 194 pages.

Lynch, K. (1981). *Good City Form*, The MIT Press, Cambridge, Massachusetts, 514 pages.

Mitchell, R.B. and Rapkin, C. (1954). *Urban Traffic: A Function of Land Use*, Columbia University Press, New York, 226 pages.

Monheim, R. (1988). Pedestrian zones in West Germany - The dynamic development of an effective instrument to enliven the city centre, In Hass-Klau, C. (editor). *New Life for City Centres: Planning Transport and Conservation in British and German Cities*, Anglo-German Foundation, London, pages 107-130.

Naess, P. (1993a). Energy use for transport in 22 Nordic towns, NIBR Report No 2, Norwegian Institute for Urban and Regional Research, Oslo.

Naess, P. (1993b). Transportation energy in Swedish towns and regions. *Scandinavian Housing and Planning Research*, Vol 10, pages 187-206.

Newman, P. (1990). The search for the good city. *Town and Country Planning,* Vol 59, No 10, pages 272-275.

Newman, P. and Jennings, I. (2008). *Cities as sustainable ecosystems: Principles and practices*, Island Press, Washington DC, 296 pages

Newman, P. W. G. and Kenworthy, J. R. (1984). "The use and abuse of driving cycle research: Clarifying the relationship between traffic congestion, energy and emissions". *Transportation Quarterly,* Vol 38, No 4, pages 615-635.

Newman, P. W. G. and Kenworthy, J. R. (1988). "The transport energy trade-off: Fuel-efficient traffic versus fuel-efficient cities". *Transportation Research,* Vol 22A, No 3, pages 163-174.

Newman, P.W.G. and Kenworthy, J.R. (1989). *Cities and Automobile Dependence: An International Sourcebook,* Gower, Aldershot, England, 388 pages.

Newman, P.W.G. and Kenworthy, J.R. (1999). *Sustainability and Cities: Overcoming Automobile Dependence,* Island Press, Washington DC, 442 pages.

Peirce, N. (2009). German city emerges as a world class energy saver, http://citiwire.net/post/973/ Washington Post Writers Group.

Punter, J. (2003). *The Vancouver achievement: Urban planning and design,* UBC Press, Vancouver, 447 pages.

Putnam, R. (2001). *Bowling Alone: The Collapse and Revival of American Community,* Simon and Schuster, New York, 544 pages.

Rees, W.E., Wackernagel, M. and Testemale, P. (1998). *Our ecological footprint: Reducing human impact on the earth,* New Society Publishers, Gabriola Island, BC, 160 pages.

Salomon, D. (2009). Freiburg Green City: Approaches to Sustainability. Presentation to European Green Capital Award, 12.01.2009. http://ec.europa.eu/environment/European greencapital/docs/cities/20102011/freiburg_presentation.pdf

TEST (1989). *Quality streets: How traditional urban centres benefit from traffic calming,* Transport and Environment Studies, London, May.

Tri-Met (2009). *Status Report*, Spring. 20pp.

Watt, K.E.F. and Ayres, C. (1974). "Urban land use patterns and transportation energy cost", Presented to the Annual Meeting of the American Association for the Advancement of Science, San Francisco.

In: Eco-City and Green Community
Editor: Zhenghong Tang

ISBN: 978-1-60876-811-0
© 2010 Nova Science Publishers, Inc.

Chapter 4

COMPARISON OF PLANNING THE SUSTAINABLE ECO-CITY IN READING AND TAIPEI

Szu-Li Sun

University of Reading Business School, England

INTRODUCTION

Both Taipei City in Taiwan and Reading town in England have put significant efforts on policy-making and implementation aimed at developing a sustainable Eco-city. This paper aims to explore how the sustainable Eco-city policy was deployed and interpreted at a local city level in different contexts (political, cultural and economic), with particular emphasis on land use and transportation. The research strategy involves cross-national comparison. It discusses and compares the local and global contexts of the two empirical studies to help understand how sustainable eco-city policy is deployed at a local city level. It investigates the issues surrounding policy formulation, the structure of the planning process and the outcomes of over a decade of (variable) policy implementation.

The evaluation is based on a theoretical propositions of the sustainable eco-city concept with four principles (environment, futurity, equality and participation), developed from Szu-Li Sun's PhD thesis (Sun, 2008). It is used to investigate the evidence, although local interpretation of the principles is acknowledged and incorporated. In order to collect the evidence, the interviews in the two cases were conducted. As a result, the discussion will identify some of the vital factors which influence the planning process and outcomes.

Taipei Case

The first case study analyses the local administration of Eco-city policy in Taipei, which has been promoted by Taipei City Government. The emphasis here has been on environmental protection but it has also encouraged planners to involve the community in aspects of urban design. The starting point for exploring the concept and planning process in

Taipei City must therefore be to examine the background and policy evolution that urban areas make to sustainable economic, social and physical development, both locally and globally.

Taipei City is located in the Taipei Basin in the northern part of the island of Taiwan in the centre of East Asia. Taipei is the capital of Taiwan and also the political, economic, educational and recreational centre of a country consisting of 33 cities and counties. The city's boundary, which is 27,179.9 hectares in area, is encircled by Taipei County, itself comprising 205,257 hectares. Moreover, the close relationships of Taipei City, Taipei County and some of Tao-Yuan County have become a Great Taipei Metropolitan area. It also made urban problems more complicated and these cannot be solved by Taipei City itself. Around Taipei City and including Yuan-Ming Mountain National Park, are mountains while meandering around the city are three rivers, the Danshuei River, Jilong River, and Jingmei River. The entire city is nearly fully developed except for some mountains and dangerous flood plains which are under enormous pressure to be developed because the average population density is very high at 9,684 persons per square kilometre with a population of 2,616,520 (11.4% of the total population of 23,037,031 in Taiwan in 2009).

Taipei City is located in a fragile environment, a dead volcano becoming a flood sludge basin. Some disastrous landslides caused by development on mountain slopes, that killed several people reduced pressure to develop a decade ago. In order to prevent flooding, the rivers were lined with higher and higher embankments which have already been elevated by more than 9 metres. Even though on top of the embankments a few bicycle paths and parks were built recently, people still cannot easily reach the river. Recently calls for the right of access to the river were emerging and there was some debate during the interviews. However, before the flood problem could be solved, no government officials were willing to lower or demolish the embankments. For example, when facing the Taipei City councillors' strong enquiries, Taipei City's ex-Mayor Ma claimed that if anyone can guarantee that there would be no flood in the future, he would agree to demolish the embankments. This is not good practice for an eco-city.

Due to the Central government's strong top-down-policy emphasis on economic development with little regard for environmental considerations, the economic growth has been described as a miracle and as one of four East Asia "dragon" economies. Taiwan's economy depends upon imports and exports. The economy was increasingly strengthened but with an increasingly centralised political regime and little consideration of environment. Economic competitiveness has also been a central concern of the Taiwan government. This is particularly so in relation to neighbours countries in Asia, especially mainland China.

Reading Case

Reading case study explores a Green City policy in Reading with a focus on green transport and green economy. The planning process involved significant inputs from a community network. The planning policy framework for Reading's development has been influenced by a number of inter-related forces, which includes the national and local political context, economic pressures, and a re-orientation in focus and approach provided by the new UK planning system.

Reading is strategically located in the centre of the county of Berkshire, playing a vital role in the south east region - especially in the Thames Valley. It is 40 miles west of London, and near Heathrow Airport (both accessible via the M4 motorway). Reading Borough covers 4,040 hectares (= 40.4 sq. km.) and is comprised of 16 voting neighbourhoods (wards). It is one of the principle regional and commercial centres of the Thames Valley, and the second largest interchange station on the UK rail network (Reading Borough Council, 2003, p. 3). Reading is the hub of the Thames Valley and has some of the most significant high-tech business parks in the South East. It has changed from a middle sized industrial town to a key regional centre (HM Treasury, 2005, p. 1) and it is the second largest office centre in the south east region of the UK outside London (Breheny and Hart, 1994, p. 1). How does this strategic location and the buoyout economy affect Reading's eco-city project?

Reading's economic success and prosperous development have created some characteristics and problems in the planning process toward a sustainable Eco-city.

The case study evaluation starts from acknowledging the perceptions of local authorities and local residents, and then explores the process of local policy making and implementation. By exploring these, some critical and influential factors are identified.

PERCEPTION OF THE SUSTAINABLE ECO-CITY

This section explores the context and perception of a sustainable eco-city policy to show the effect of political, economic, institutional, geographical and cultural contexts, both locally and globally.

First, it was found that local and global contexts can have a strong influence on the perception and policy evolution of a sustainable eco-city. Many factors in both local and global contexts are intertwined and interdependent. For example, the unstable international relationship with mainland China may affect Taiwan's local politics and economic situation. The economic context also affects policy making to do with the environment. Similarly, the EU has had an impact on the consideration of Reading's environmental and economic issues, devolving a range of eco-city concepts, policy guidance, regulations and some funding down to 'structure' local decision-making. Secondly, the perception of the sustainable eco-city was interpreted depending on the practitioners' knowledge and involvement in local and global contexts. For instance, some academics and local government staff had a clearer, more technical, understanding of the key principles and requirements of a sustainable eco-city than local community representatives because they had more experience in the shaping of green city policies such as Local Agenda 21 and Reading 2020.

There are some similarities between the two case studies. Both have gradually put more effort into planning for the sustainable eco-city by facing countless economic and development pressures. Other similarities include the social/cultural context, institutional and organisational arrangements, and economic and political factors. For example, the political context has played a vital role in guiding the policy making in both studies. Moreover, it is apparent that economic competitiveness has influenced Taipei and Reading's path towards the sustainable eco-city and it appears significantly stronger in both cases than the environmental principle in sustainable development.

LOCAL AND GLOBAL CONTEXT

In terms of local and international context, both case study cities have rapidly growing economies with high technological industry and their main goal is to pursue economic competitiveness in response to globalisation. This has some impact on their policy making; and sometimes very clearly at the expense of the environment. Recently, both countries have made some progress in relation to their indicators of economic and environmental 'success' (see Table 4a). In both cases, economic globalisation provides a new challenge and opportunity for local government to cultivate economic growth whilst, at the same time, taking into account social equity and environmental issues.

Furthermore, local and international politics have influenced the sustainable eco-city policies in both cases. The strategies in both Reading and Taipei have been strongly affected by party political ideologies and debates and, in particular, the "greener" eco-city concept has been largely derived and forged from the more socialist-orientated parties; the Labour Party in Reading and the Democratic Progressive Party (DPP) in Taiwan. Reading has been controlled by the Labour Party for 16 years and the city's policy history has been heavily influenced by the political debates arising from this source. Overall, the political thought of the Labour Party has been quite closely linked to the four principles of the sustainable eco-city and this has helped the flow of such ideas into most of Reading's relevant policy documents. However, economic factors have also played a key role in shaping much of the policy making. Therefore, the attitude of Reading Borough Council (RBC) has often involved a compromise in which arguments for 'balancing' the environment, society and the economy have usually been used.

What is more, the EU has had some influence on UK environmental policy and funding and its interest in planning and environment has increased significantly over the last three or four decades. During the early period it operated in a more top-down manner but more recently it has developed a bottom-up or mixed pattern of influence, in which policies and procedures have been more 'democratic'. This EU influence has generally encouraged Reading stakeholders to walk closer to the path of the sustainable eco-city, at least the version promoted by the EU.

In Taipei, after DPP City Mayor won the election and introduced the four principles into government documents the Kuomingtang Party (KMT) began to pay attention to them and include them in documents of their own. Although only holding power for 4 years, the DPP City Mayor adopted a reformist approach which has made significant changes to policy making. The unstable political and administrative contexts in Taiwan prevented the Taipei City Government (TCG) from having a very effective tax-base and this hampered it from being able to implement many reforms or new initiatives. Unsurprisingly, international politics has played a key role in 'structuring' the Taipei case, mainly as a consequence of the volatile relationship with mainland China. Moreover, due to historical factors, it is easy to detect the various influences from Japan and the USA in the Taipei City urban plans.

Political culture has significantly influenced policy making in both Reading and Taipei. However, it is partly different because of the distinct regimes operating in each city. RBC is a councillor based regime, whilst Taipei City uses a mayoral approach. RBC's members are elected in sixteen wards and the winning party takes overall responsibility to lead and manage the council's affairs. They also have to prepare bids and plans to secure financial support from central government and the EU. The Taipei City Mayor is elected by the whole city and leads the Taipei City Government (TCG), while Taipei City councillors elected in twelve districts supervise the performance of TCG. Reading has had problems in recruiting councillors devoted to local government due to poor pay, heavy work commitments and limited power and influence. However, this has also meant that there are few corruption opportunities. Despite the problems with recruitment, they appear to provide capable leadership and have been moderately successful in leading the council towards a green city objective.

On the other hand, with high pay and relatively wide powers, the performance of Taipei City councillors has not been very productive in driving the City along the path towards a sustainable eco-city. For example, one of the most significant political cultures pertains to most of the councillors enthusiastically requesting development such as the construction of roads, buildings and facilities, for their supporters. Some of them might even be regarded as obstacles on the path to a sustainable eco-city because they subscribe to the doctrine "Not In My Back Yard" (NIMBY) and political corruption casts a shadow over all areas of decision-making. Fortunately, younger councillors have recently brought environmental and participation ideals into the city's decision making and they can be seen as the more progressive forces towards the development of the sustainable eco-city model in Taipei.

Cultural context also plays a vital role in relation to both case studies, with certain aspects of western culture in Reading, and similar elements of eastern culture in Taipei helping to structure both the planning and development processes. In Reading, a strong 'consumer-oriented' culture and the lack of tough social sanctions against littering for example, have affected policy-making and implementation in both transport and land use programmes. It has led to policy compromise and ineffectiveness and to an outcome not as significant or environmentally sustainable as expected. On the other hand, though eastern culture such as the Chinese style of autocratic monarchy dominates the Taipei approach, a number of foreign influences have had some significant impact in Taipei City, seen in for example, a creeping emphasis on public participation and modernistic architectural design, living style and transport policy. Currently the City looks more like a western city with a lot of westernised townscape features.

Table 4a. Environmental and Economic Indexes for the UK and Taiwan

Index	Country	2001	2002	2003	2004	2005	2006
Environmental Sustainability Index ranking	UK	16	91	-	-	65	5
	Taiwan	-	-	-	-	145	24
Growth Competitiveness Index ranking	UK	12	11	15	11	13	10
	Taiwan	7	6	5	4	5	13

Source: 1. The Environmental Sustainability Index report, 2001, 2002, 2004, 2005 and 2006. 2. The Growth Competitiveness Report, 2001-2006.

In addition, particular geographical factors have affected policy making. Both case study cities are located at strategic intersections in the global economy and, although they each have limited natural resources, they have nurtured high technology industry using their strategic transport locations. This has allowed their local economies to flourish, which has maintained pressure on the environmental objectives of the local sustainable eco-city policy.

The evidence in the two cases suggests that economic, political, institutional, cultural and geographical factors have significantly influenced the policy making and implementation processes. Furthermore, these factors tend to be intermingled so that economics influences politics and institutional arrangements which then 'feedback' into cultural understanding and characteristics. Releasing these forces is never easy and we have to leave some elements of unpredictability in the inter-relationships at play. Furthermore, all these processes and influences are the product of (constrained) human choices, which can invent new ways of working over time. Indeed, to some extent that is what the Sustainable Eco-City ideas are doing in both of these case studies. It is, therefore, the actions and perceptions of key individuals to which we now turn in order to help us understand this human dimension of policy making and implementation process.

THE PERCEPTION OF SUSTAINABLE ECO-CITY IDEAS BY PRACTITIONERS

The empirical evidence shows that interviewees have different levels of understanding in articulating and defining their perceptions of a sustainable eco-city. They were also affected by the local and international context as well as their own knowledge and degree of involvement.

Local and International Contexts

The local and international contexts played a main role in affecting the perception of a sustainable eco-city. For example, perceptions of the eco-city in Taipei are evolving differently according to changing economic conditions in particular. In Taipei the sustainable eco-city was often interpreted as meaning an environmental city and was generally narrower in definition than in Reading. This was partly due to the economic transformation process. Currently in Taipei, to cope with the trend of sustainable development, the term 'sustainable city' is used instead of the term eco-city, which is seen as out of date and too heavily linked to an 'environmental' perspective. It results in the policy making and implementation of eco-city policy in Taipei being focused more on environmental issues.

On the other hand, as the stakeholders of Reading have long been involved in the Local Agenda 21 process, it means that local people have a better understanding of the meaning of a sustainable eco-city. Therefore, in interviews, the key players in Reading use definitions more related to the four principles of the proposed analytic framework.

Practitioners' Knowledge and Involvement

It is proposed that practitioners have different perceptions about the definition of a sustainable eco-city on the basis of their individual knowledge and involvement. Such an assumption can be supported by the empirical evidence. Some local government officers and academics had a clearer, more technical understanding of the key principles and requirements of an eco-city than the local community representatives. A practical understanding is achieved through their involvement in the planning process leading to greater understanding of the perspectives about the key principles.

For example, in the case of Reading, the perceptions of a sustainable eco-city were expressed by most of the interviewees in terms of sustainable development, sustainable transport, protection of open space, recycling, eco-footprints and participation. Although some of the community representatives could not directly articulate this view, some of their interpretations were clearly linked to these concepts. They made suggestions which were closely related to the environment, futurity and participation principles. Only a few argued for the inclusion of equality principles. This perception also affects the way that policy and implementation have been developed, as we will see later.

On the other hand, in terms of the three themes (the economy, society and the environment) most of the government officers conceptualised economic growth as the solution to most problems in both cases. Their idea of balance between these three themes tends to marginalise the social and environment problems, as these two are continually constrained by the need for economic development or growth. Therefore, the same attitude continually reflects on policy making and implementation as well.

THE PLANNING PROCESS OF THE SUSTAINABLE ECO-CITY

The discussion will examine how planning for the sustainable eco-city is carried out at the local level. Some findings from the empirical evidence will be discussed in relation to the following issues:

Planning Trends

1. Formulation and Evolution of Eco-City Policy in Relation to the Four Principles

The empirical studies demonstrate that the formulation of an eco-city policy is, to some extent, an evolutionary process. In Reading, for instance, there can be found quite strong environmental protection policies as far back as the 1951 Local Plan. This document's policies for the redevelopment of existing congested residential areas show that the planners were aware of the importance of the environment and futurity principles. The council has continued to preserve some open spaces and restrict development to some main areas such as the town centre.

In Taipei, the review of eco-city policies and documents can be traced back to the first plan drafted during the Japanese occupation of 1895-1940. At that time the plan focused on transportation and park design, aimed at relieving the heat of Taipei City. During the 1960s and '70s the slogan *"urban ruralised and rural urbanised"* was promoted by the Taiwanese government as the concept for urban development. This concept represented the Chinese political leaders' acceptance of the western garden city planning idea which they then introduced into the land use guidance for town and country. The model had similar elements to the eco-city concept.

Over the last 60 years, the four principles of sustainable development were gradually embedded in the green city policy in Reading and the sustainable eco-city policy in Taipei.

2. Key Factors Influential during the Sustainable Eco-City Policy Evolution

During the period of policy formation and evolution, some key influential forces can be identified. These include historic, economic, cultural, geographical, institutional and political factors.

In terms of government institutions, some key relationships involving tensions, competitiveness and cooperation between local and central governments have had a significant influence on both cases' planning process. The philosophy of competitive bidding existed in both case-studies though within different resources and government institutions. With limited resources and those mainly controlled by central government, both case-study local authorities had to develop their own unique and creative plan to gain financial support to implement their eco-city policies.

The performance of RBC, for example, is assessed by central government through 'best value' indicators. With limited resources and lack of staff the council has to compete with other local authorities to gain financial support and attract investment. The competitive bidding approach exists in Taipei as well. The sixteen districts have to compete just in order to apply to TCG for financial support. The same situation exists for the TCG which has to compete with twenty three other local authorities to gain central government financial support. It is an efficient mechanism for controlling government funding and encouraging competitiveness among local authorities to stimulate innovation towards a sustainable eco-city. Centralism and economic competition provide some constraints on Reading and Taipei's freedom of manoeuvre and drives their eco-city policy in ways that tend to emphasise the economic dimensions of sustainable development.

Another significant factor influencing the eco-city policies of both cases was local politics. In the Reading case, the planning process was strongly affected by the political ideas of the Labour Party and involved participatory processes, but it also needed institutional change to help in delivering the principles of inclusion and equality. In Taipei's case, the effect on planning was due to the historical influence of Japan and the USA, plus the KMT and DPP Parties. Moreover, the Chinese political culture of autocratic monarchy has slowed Taipei City's progress towards a sustainable eco-city, but the Western tradition of democratic politics is now broadly accepted. Whatever weaknesses this model has (for example, Taipei City focuses on a technology-based approach and vehicle-centred set of policies), its influence on local politics in Taiwan has generally been progressive in terms of the participation principle of eco-city policy.

New Labour talks of 'revolutionising' and 'modernising', but in reality it practices evolution within a capitalist economic framework. In this sense New Labour is not as radical as Old Labour, but at least it has committed itself to the concept and strategies of sustainable development.

In both cases, it is evident that being an internationally competitive city has become the most important policy goal, following processes of globalisation. This has provided a significant constraint on the radicalism of the eco-city strategies that could be developed in the two studies, with a constant requirement to moderate environmental priorities according to the needs of economic 'balance' (Reading) or international competitiveness (Taipei).

3. Long-Term Trend of Planning Systems towards a Sustainable Eco-City

In both case studies, it was found that the planning system has accommodated the four principles of a sustainable eco-city policy. For instance, participation has been initiated, but not in a radical way (to fit concepts of 'empowerment', 'self-management', etc.).

In the case of Reading, the new planning development framework introduced in 2004 has more overtly embraced the participation, equality, futurity and environment principles. The New Labour government has committed itself to encouraging participation, especially based on the new system in 2004. They have encouraged greater public participation in the planning system, but also sought to speed-up decision-making, indicating the constant tension that affects these areas of public policy. However, they have also emphasised the role of regional planning bodies and local planning authorities in exercising their functions with the objective of contributing to the achievement of sustainable development. The new system requires local government to prepare sustainability appraisal reports for their local development plans. It attempts to help planning authorities fulfil the objective of contributing to the achievement of sustainable development in preparing their plans. The implementation of the plan should, therefore, help to achieve the social, environmental and economic objectives.

Furthermore, four significant mechanisms (i.e. Reading's Local Agenda 21, the Community Strategy, Reading 2020 Partnership and the Local Development Framework) are designed to make significant contributions to the planning process. These approaches have established community networks to help formulate policy and, in doing so, they have embraced the participation principle, moving from local government to local governance. In some respects these mechanisms have contributed to RBC's process of embracing empowerment, self-management and decentralisation concepts.

In Taipei City, in order to help empower the local community, the power to approve local plans was transferred in 2002 to local authorities by central government. Because central government realised decentralisation was necessary and unavoidable, they gave local authorities more responsibility for urban planning helping the planning system to make a more effective contribution towards the goals of the sustainable eco-city.

In addition, the community planner system has been set up and has successfully created bridges between the community and government where planners help residents and disadvantaged groups to conduct environmental reviews and make suggestions of improvements to public open spaces. Unfortunately, the community planner system did not connect with the urban planning system as closely as it could have.

In both Taipei and Reading, the planning system has changed to embrace the four principles to some extent, though not to the same degree in both cities. The participation principle especially has recently been strongly highlighted in both cases with regard to citizenship and partnership. The planning system is gradually changing to shape the future of places through empowerment and creative mechanisms such as strategic planning and partnership.

4. Aspects of Centralisation and Decentralisation

The main characteristics of the planning process of the sustainable eco-city approach should include self-management, decentralization, empowerment and participation.

Therefore, part of the evaluation in the case study involves examining the planning process to see if self-management, decentralisation, empowerment and participation are embedded in it. Both cases adopted a top-down planning approach, centrally controlled but recently moving towards a more decentralised system. This same situation existed in both the relationships between central government and local government, and between local government and local residents.

It is evident that in Reading the planning process appears more decentralised and empowered (to local communities) and the centralised decentralisation intention of central government is clearly seen. For example, Reading's Local Agenda 21, the Community Strategy, Reading 2020 Partnership and the Local Development Framework are designed to empower local authorities and communities. Though some problems still remain, such as NGO and community claims that it does not empower them enough, generally speaking, the decentralisation has already been embraced by central government and RBC.

In Taipei, it is mainly a government-led regime and there was little real public participation in the urban planning process. Taipei has produced its 'Community Planner System' to improve citizen participation, encouraging planners, architects and citizens to join in the 'Neighbourhood Improvement Programme' and asking for the community and the professionals to form a partnership. Both central and local governments used to operate through a strong centralised regime, and there is a concern (by government officials and politicians) that recent attempts towards decentralisation will result in a reduction of government power, both central and local.

5. Movements from Local Government to Local Governance

The transformation of local government to local governance reflects the intention of central government to decentralise responsibility and empower local government and local residents, albeit within a framework of controls and targets. The shift to local participative governance in which the local community and their representatives play a much more active role in urban projects, is illustrated in the regeneration process.

The Reading case provides an illustration of how the transformation of local government to local governance in the planning process appears to be moving significantly towards a green city policy. With strong traditional community involvement, Reading LA21 and the Local Strategic Partnership have played key roles in facilitating the transformation process. The planning process of Reading has developed from state control and regulation of urban policy, towards local empowerment and decentralised decision-making. The establishment of the Reading 2020 Partnership shows how the Labour government's emphasis on partnership has been operationalised within a framework that supports "centralised decentralisation". It is a new form of governance that has emerged where the state has permitted more communities to implement programmes and take greater responsibility for their own development.

This decentralisation of decision-making to the local community is clearly in line with the participation principle of an eco-city policy, but sceptics maintain that it merely represents government "at a distance". Community representatives and voluntary organisations believe that decisions have been made before public participation; there were invisible controls in the decision-making process that operated before committee and cabinet meetings. They argue that informal participation is greater than formal participation.

PLANNING MECHANISMS

During the planning process, some findings regarding particular planning mechanisms can be identified in the empirical evidence.

1. Community Participation

The democratic basis of participation is for greater community involvement. In both cases local authorities have recognised it and attempted to improve community participation. The two cases provide evidence to identify the discourse among stakeholders involved in the sustainable eco-city policy. The discourse is explored not only through top-down processes but also via bottom-up ones, in which it becomes embedded within the institutional structure and practice. Reading and Taipei have attempted to set public participation within their relevant institutional structures using community strategies and local strategic partnerships (in the UK case) and the community planner system (in the Taiwanese case).

However, it became apparent during the interviews, that the mere existence of some official discourse advocating empowerment and partnership does not guarantee that it will actually be translated into practice or that the intention of such a discourse is genuinely to empower communities through participation in partnerships on urban projects. The interviews explored how partnership and empowerment have been conceptualised by different stakeholders. In Reading, community representatives and voluntary organisations all appear distrustful and sceptical of government officials.

The case studies provide evidence that community participation can enhance the effectiveness of regeneration programmes in terms of better decision-making, more effective programme delivery and sustainability of regeneration programmes. Examples of this include the 'success' of Reading' town centre regeneration. However, the movement towards greater and more effective community participation still has some difficulties. For example, both cases show that one of the problems is that it is a typically government and developer-led urban planning system, with only limited opportunities for non-elite interests to shape it, other than by simply objecting to developer or government initiated proposals.

Moreover, both cases imply that complete empowerment of communities is currently very difficult if not impossible, as the breadth of community representation is, as yet, inadequate. Therefore, key challenges facing both countries are how to encourage the full range of stakeholder groups in the local community to participate and how to resolve the inevitable conflicts of interest that arise. Both cases show that the local authorities have attempted to go beyond rhetoric by taking the issue of community participation seriously and improve the extent of community involvement a little, but there are still problems. Though it offers a new opportunity to let disadvantaged groups have a say in the revitalisation of communities, the ideal state 'required' by the eco-city model has not yet been fully achieved.

2. Community Network and Consensus-Building

One of the significant planning mechanisms is to build community networks at the local level. Both case studies show that, with local government's help, the community networks are one of the most important instruments to bring together local communities, stakeholders and partners and formalise a range of community strategies. This planning mechanism provides an important tool in the process of consensus building.

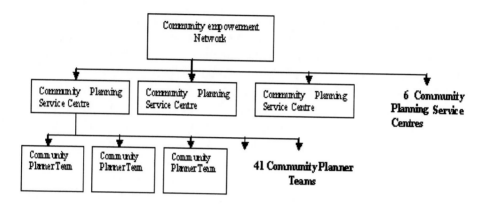

Figure 4e. The Community Network which Taipei City Government Proposed.

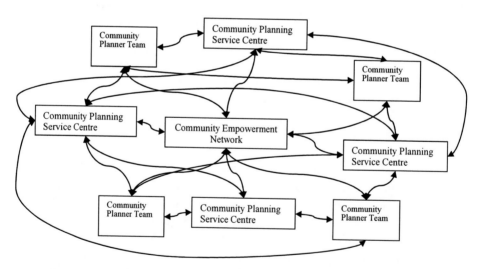

Figure 4f. The Community Network which the Community Empowerment Network Proposed.

For example, 'Reading 2020 – Making it Happen' is Reading's first community strategy developed by Reading's Local Strategic Partnership to *"improve the quality of life for our communities by responding to local needs and reflecting national priorities"* (Reading LSP 2003, p1). This partnership approach has strengthened the community network in Reading and the resulting planning process has focused on equality and participation principles. Through the community network, different stakeholders have been given opportunities to communicate and participate in building the strategy for the area which seeks to deliver a sustainable future.

Taipei also produced a quite successful community network through the community planner system (established in 1999) and Community Empowerment Network Programme (established in 2004). However, there were different arguments about building the community network between NGOs and TCG staff. TCG planners believed that the community network was like a tree hierarchy in which the Community Empowerment Network was on the top. The interviews painted a picture similar to that depicted in Figure 4e. However, the Community Empowerment Network planners argued that the network should be drawn like a web, as in Figure 4f. They believed all the agencies and stakeholders were nodes and they could help all the nodes to connect and share all the resources.

The community network establishment provides a consensus-building forum and bridge so that all stakeholders can communicate with and participate in government projects. The two cases show that networks can facilitate community participation. Reading developed a community network much earlier than Taipei, so some of the participation experiences could provide a good lesson for Taipei. As TCG has only just started to encourage community participation, the recent initiatives could be regarded as the 'green shoots' of democratic participation. Both cases realise that participation is one of the key planning mechanisms necessary to achieve a sustainable eco-city. Building community networks at the local level to provide consensus-building is the practice of the participation principle. However, both cases show some evidence that local authorities do not empower the citizen enough as they are afraid that their power and influence will be taken away if real citizen participation is introduced.

3. The Dynamics of Power and Influence

The dynamics of power, influence and resistance were evident in the relationships between the stakeholders involved in the eco-city planning process. They included central and local government, community and voluntary groups, private-sector organisations and regional government-funded agencies.

The analysis of the case studies shows that power is one of the key underpinning resources for empowerment. For example, the Taipei case provides clear evidence that local government has attempted to encourage community participation but it has been rather cautious when 'delegating' resources and decision-making responsibilities to local communities. Though it set up the Community Planner System, it appears to be more symbolic than a committed attempt to 'empower' local citizens.

Similarly, it is evident that central government is afraid of genuinely empowering the TCG. There are some reasons for this. For example, since 1996 leadership of the TCG and central government have been taken by different political parties. They compete and have very different, often opposite stances on many issues. This party political conflict has strengthened the desire of central government to put constraints on the City Council and affected movement towards local empowerment.

Formal party political power also plays an influential role within the eco-city policy. For example, in Reading, although some of the council staff and councillors are aware of the importance of citizenship and new localism in building a participatory approach, they are also worried that their own political influence would be eroded by the increased role of local environmental groups, like GLOBE. In particular, the Labour councillors are afraid of losing their leadership role in Reading and have been keen to hang-on to representative democratic power. On the other hand, local Liberal Democrats and Conservatives have encouraged the increased opportunities for 'active citizenship' in order to increase their own political influence.

Another key issue is the influence of local businesses, voluntary organisations and minority ethnic groups in relation to local government. For example, in Reading interviewees from the voluntary sector argued that the business sector has gained a privileged position in communicating with the council because the council has committed itself to an economic growth strategy. At the other end of the spectrum, the minority ethnic groups feel powerless and indifferent to local affairs. This power-imbalance between key stakeholder groups has a variable, but sometimes significant influence on the eco-city planning process. For instance, the RBC is keen to accept local business's opinions to advocate the development of the Kennet Valley area and the Hills Meadow theatre project despite the disagreement of many local residents and environmentalists.

Arising from this conclusion, it can be argued that it is necessary to seek out more efficient and effective ways to secure a more active role for local community representatives in order to gain the confidence needed to challenge the position of the "powerful" stakeholders, such as business.

4. Participatory and Representative Democracy

The analysis of the case studies provides detailed examples demonstrating the types of democracy that have been adopted in the planning process for the sustainable eco-city. In both cases, a combination of representative and participatory strategies can be seen.

In Reading citizens have been allowed to participate and express their opinions in direct ways (e.g LA 21, Reading Community Strategy and consultation), but the elected councillors have still made the final decisions on fundamental issues of policy and, to some extent, implementation. The council is "greening" its policies and practices and this has arisen through pressures coming from the "participatory democratic input" of the GLOBE (and other local) groups along with "the representative" input of elected members (Doak, 1998, p. 87).

Taipei City might be regarded as a firmer (old-fashioned) representative democracy but it is also moving slowly towards a more participatory form of governance. Because of the historical political factors, Taipei democracy has developed very late. In the interviews conducted for the research, one of the TCG staff argued that they had provided the path for greater direct citizen participation, but most other interviewees said that there were too limited opportunities to participate. Some even argued there was no real democracy at all. This was illustrated through the top-down institutional design of government and the evidence of elite participation during the policy making process.

5. The Adoption and Significance of the 'Third Way' Approach

In terms of the planning approach, both Taipei and Reading seem to have taken a similar path (which can be broadly called the 'Third Way'), although it is increasingly being revised and aspects of centralism still tend to dominate. The comparison of the Taipei and Reading current approaches to sustainable eco-city principles can be seen in Table 4b.

In Reading, the council has taken on board the Third Way approach in the planning process, as RBC has been controlled by the Labour Party for sixteen years. New Labour's programmes of reform have been introduced and absorbed to try and produce an integrated approach to deal with urban affairs. New Labour has tried to promote a holistic, cross-sectoral approach on sustainable development since 1997, though it has only achieved it to a limited extent. *"Holistic" is a term to describe "developing a cross-sectoral approach on issues involving different departments and agencies"* (Driver and Martell, 2000, p. 149). New Labour is gradually reforming local government along these lines, taking a compromise position and not dramatically changing the organisation of local government at each stage. Tony Blair's Third Way approach is based on free-market principles *"by having a laissez-faire view of the state"* (Driver and Martell, 2000, p. 149). The New Labour approach is solidly within a compromising centre-left Third Way rather than grown from the Party's historic socialist roots.

In Taipei City, Mayor Ma has also promoted a so-called Third Way approach (Hsiao Hsu-Tsen, 2005), similar to Tony Blair's ideology. As the KMT lost the Taipei City Mayor election in 1993 for the first time, this experience gave KMT a chance for self-examination after a long period of control. Ma's success at KMT's heart in the TCG was partly because of his tremendous leadership and personality, and partly because of the former DPP mayor's reformist approach which was not easily accepted by large sections of the public. Therefore, the current mayor has taken a more compromising approach while continually revising it.

6. Training and Capacity-Building for the Community

The two case studies provide evidence to show that efficient and effective citizen participation needs the development of appropriate knowledge and skills. It is necessary to build the confidence, skill and knowledge of individuals and community organisations in order for them to take-up the opportunities for participation and 'empowerment'.

In Reading, even though the training of community groups and representatives is very important to community participation, RBC was doubtful that the government was capable of giving its citizens such a training course. One of the council staff expressed his many doubts during interviews: *"Is the government the best agency to give a training course for the citizen? Does the government have the staff and skill to train the citizen? Does the government train the citizen to say what we want them to say?"* He also complained that, without practical experience, central government did not realise how difficult it was to build effective public participation.

The Taipei case shows that it is important to engage in a process of training and capacity-building for the community in general and more specifically for community representatives. For example, in Taipei City if the leader of the Li administration had taken part in the training for the community planner system, he would have better understood how to propose a community development project for the local area and be able to apply for financial support to the TCG.

Table 4b. The Comparison of Taipei and Reading's Current Approaches to Sustainable Eco-City Principles

Area	Taipei		Reading	
	Problem	Solution	Problem	Solution
Approach	"Third way" Compromised Centralism	Highlight Decentralisation	Third Way (the central left) Greater supply-side emphasis within strict monetarist approach	Highlight Decentra-lisation
Environ-ment	Environment =Economy Technocratic resolutions Anthropocentric	Eco-centric approach Ecological modernisation	Environment= Economy Technocratic resolutions Anthropocentric	Eco-centric approach Ecological moderni-sation
Futurity	Middle emphasis on futurity	Highlight futurity Internationalism Sustainable development	Middle emphasis on futurity	Highlight futurity Interna-tionalism Sustainable development
Equality	Low emphasis on equality	Highlight the need of disadvantaged groups	Middle emphasis on equality	Highlight diversity Inclusion (Both Rights and Responsi-bility) Social justice
Partici-pation	Elitism Citizenship	Empowerment Citizen Participation	Individual and community focus Citizenship Partnership New localism	Empower-ment Citizen partici-pation

7. Highlighting the Equality Principle

The analysis of two case studies shows how the equality principle was given much more emphasis than before within the planning process.

In the Reading case, RBC followed the path set by New Labour. Two of the Labour Party's core ideologies, equality and democracy, seem similar to two of the sustainable eco-city analytical principles, equality and participation, but have a slightly different transformation. For the Labour Party, equality is used to imply reorganising society with the specific object of creating a more equal distribution of wealth and power, not just for individuals. This has moved towards a softer stance in which equity issues need to be considered, but the reformist ideals have been largely side-lined. Therefore, the equity issue is changing, shifting from wealth distribution advocated by Old Labour to wealth creation advocated by New Labour. New Labour seems to walk closer to Conservative thought on this issue. However, the implication for the planning process is that they emphasised equality, social inclusion, cultural diversity and citizenship.

In Taipei, the DPP introduced a strong(ish) element of socialist and reformist thought into TCG. In doing this, the equality principle was highlighted and policy became more concerned with the needs of disadvantaged groups. The DPP Party took its support mainly from social movements, such as those supporting the environment, aboriginals, farmers and organised labour. After constituting a political party around these groups, the DPP transformed the social movement ideals into social policies. The National Health Insurance, Unemployment Insurance and National Annual Pension have been incorporated into policy and implemented due to their efforts. The DPP acted as a stimulus to trigger Taipei to pay more attention to the equality principle. They have put more emphasis on policies for disadvantaged groups such as people with disabilities, women, aboriginals and young people.

The analysis of the planning process for the sustainable eco-city in both cases provides detailed examples to illustrate the particular dynamics of the process. Some key influential factors in terms of local and international contexts were identified. The planning process is a network building process transforming political arrangements and structures from local government to local governance; from centralised to decentralised control; from representative democracy to participative democracy; gradually embracing the four principles of the sustainable eco-city.

Generally speaking, during the planning process, the different emphasis from the TCG, RBC and local residents result in different achievements. Taipei has been successful in introducing their community planner system providing a bridge between (some) citizens and local government. Meanwhile Reading has introduced their local strategic partnership aimed at improving the quality of life by responding to local needs and reflecting national priorities. In comparison, most policy making and implementation in Reading has put emphasis strongly on the participation principle which is generally lacking in Taipei.

PLANNING OUTCOMES

The purpose of this section is to discuss the empirical evidence to examine the planning outcomes of the sustainable eco-city. The analysis of the empirical study outcomes in section 5.3 and section 6.3 provides some evidence to evaluate the policy and implementation of a sustainable eco-city policy. In other words, it explores how far the concept of a sustainable eco-city has been translated into practice, focusing on the key areas of land use and transport.

Both Taipei and Reading have attempted to achieve a sustainable eco-city by producing a future vision and launching a number of significant initiatives, such as urban regeneration, transport and waste management. However, some key factors have affected them on their path towards an eco-city. There are some similarities between them. Both have gradually increased their planning efforts in the face of a number of economic and development problems. The main findings from the analysis of outcomes are as follows:

- most of the policies and visions cover some of the four principles, but to different extents;
- sustainable development indicators have been developed and used to evaluate the performance of eco-city policy;
- urban regeneration has been one of the most significant tools in progressing eco-city policy;
- green building assessment has been used to promote green design, but is still under-developed;
- green transport is potentially one of the most useful tools, but has not been very successful; and
- waste management and recycling is also a significant tool, although some issues remain.

1. Most of the Policies and Visions Cover Some of the Four Principles, but to Different Extents

Both case studies have produced vision statements targeted towards the year 2020 which are full of ambition, but which do not cover the four principles of the sustainable eco-city in all respects. In broad terms, whilst Reading has embraced all four principles, Taipei has only involved two, that of the environment and equality. This suggests that Taipei's perceptions are narrower. Though most of the policies and visions are generated from pubic consultation processes at different levels, there are clear disagreements between levels of government, NGOs and residents in both cases.

In terms of the development of these visions, the *Reading City 2020 Update,* produced in 2003 is based upon a combination of sustainable community building and the intensification of urban development. It covers the futurity, environment and equality principles but is also ambitious for economic competitiveness. Reading's development framework is designed to concentrate on the main development areas of the town centre and south west Reading. The vision and all the policies have been developed through a moderately thorough consultation process and have been subject to a form of sustainability appraisal. However despite this, the vision and policies have been criticised as politician-led documents. Political commitments to certain 'growth' areas are not without problems: the northern station area and south west Reading are located on floodplains which have generated the strongest disputes between the council, NGOs and community representatives.

The Taipei Vision 2020 encompasses only two of the four principles, environment and equality but, In fact, focuses almost entirely on international economic competitive capacity rather than environmental and equity principles. The vision involved a little public consultation and most of that was led by TCG. The development pattern and vision for the Taipei metropolitan area are designed around eight local centres, two of which are the old and new ones in Taipei City, with six transport corridors. The design is intended to relieve the development congestion in the old centre. The TCG officials believed they have produced an effective land use plan for the public, yet the NGOs and Li leader argued that historical heritage and natural resource issues were not given very much consideration.

In both case studies most policies and programme documents put the emphasis on physical development, namely infrastructure and construction, in order to build economically competitive, international cities rather than on social development. This is particularly the case in Taipei where urban development and planning history has followed the North American model. This has involved the development of massive high-rise skyscraper buildings around the old and new centres, sprawling suburbs and plans for many raised freeways and Mass Rapid Transit routes, all at the expense of heritage and low-income housing. In the case of Reading the draft Reading Central Area Action Plan (Reading Borough Council, 2006) outlines the major urban development proposed for the future of Reading. The plan takes account of the other strategic and local level plans, and maintains the emphasis on economic competitiveness by including many supporting infrastructure projects. Furthermore, central Reading will be sufficiently and intensively developed as a regional and national hub. Most of the development schemes adopt a mixed-use approach and are intended for previously developed land, in line with central government policy and certain principles of sustainable development.

On the whole, the visions and policies have been criticised as politician-led documents. Those in Reading cover more principles and elements of the sustainable eco-city than those in Taipei, partly due to more extensive and, it could be argued, intensive citizen participation. Moreover, it is apparent that economic competitiveness has guided Taipei and Reading's paths towards the sustainable eco-city and this principle is much stronger than the environmental imperative arising from the concept of sustainable development.

Thus, the policies appear to be generally in line with the principles of sustainable development but actions and implementation do not yet suggest whole hearted acceptance, as economic considerations still appear to be the primary goal for both cities.

2. Sustainable Development Indicators have been Developed and Used to Evaluate the Performance of Eco-city Policy

In both studies, sustainable development indicators have been developed to evaluate national and local performance. The local level indicators appear to better reflect the principles of a sustainable eco-city. Furthermore, it should be pointed out that the nationally-derived indicators in Reading are also designed and used to evaluate local government performance, feeding back to influence central government grants and other spending programmes. This illustrates that sustainainable development indicators can have multiple purposes (and can be constructed in different ways) and should be treated with some caution.

The Reading case provides a detailed history of the incremental development of sustainable development indicators. Reading has developed and used a great number of indicators to assess the environmental performance of RBC and the town of Reading. Though the indicators show some progress of both environmental and council performance, it was suggested that some of indicators, particularly related to participation and equality principles, could be refined to provide a more robust framework with which to evaluate Reading's progress.

In Taipei, the city government has only recently developed sustainable development indicators covering all four principles of a sustainable eco-city. The indicators proposed by TCG in 2004 have been improved and are closer to the four principles than those initiated in 1996. In particular, they cover an element of the participation principle, which the former lacked. On the other hand, when compared with the Sustainable Development Indicators for Taiwan which lacked any real consideration of the equality and participation principles, the Taipei indicators were much more firmly based on the sustainable eco-city principles.

3. Urban Regeneration Is One of the Most Significant Tools in Progressing Eco-city Policy

Both case study areas have made significant achievements in terms of urban regeneration. Both cities have used brownfield sites and opened-up the consultation process to public participation. Furthermore, they have revived the local physical environment and improved their local economies, in so far as any local authority can. Urban regeneration in both cases has put more emphasis on the economy, but in Reading it has tended to be 'property-led', using local and sub-regional partnerships, whilst in Taipei it has been more government-led without much community involvement. Yet both can be criticised for failing to significantly promote 'green economy' ideas such as eco-industrial development, green travel plans, etc..

The regeneration of Reading town centre created high-quality service-based developments and provided a focus for consumption-based, property-led economic regeneration (Raco, 2003, p. 1869). The development of the Oracle shopping centre has moved Reading up the national retail rankings from 26th to 8th in the UK. The town centre regeneration and the Oracle shopping centre have won some awards both in the UK and Europe and the council was awarded 'Beacon Council' status in 2001 for its town centre regeneration efforts.

However, there are different opinions amongst council staff, academics, NGOs and community representatives. Some of the council staff, councillors and academics argued that the town centre regeneration was in line with the environment and futurity principles as it considered the environment and the river, and provided safe pedestrianised areas for people to utilise. They argued that they were very good at using brownfield sites for urban regeneration. Moreover, the prosperous Oracle shopping centre, generated attractive trophy development which further encouraged inward investors and visitors, contributing to the town's economic success, especially in a context of increasing inter-urban competition.

On the other hand, some interviewees claimed that the Oracle shopping centre has not considered the environment and futurity principles at all. The Oracle shopping with 2,300 car parking spaces has been responsible for increased traffic congestion, petrol consumption and air pollution because it has attracted 10 million car journeys per year. Urban regeneration in Reading seems to be a reflection of the diverse ideas underpinning its green city policy, wanting to have environmental protection along with its economic growth. Though Reading has used a consensus-seeking approach, it is still open to criticism that the business sector has been given a privileged position in relation to council decision-making, largely because the council is committed to a 'sustainability scenario' based on local economic growth. This has left unresolved some important problems such as 'sustainable transportation' and the quality of the local environment (in terms of pollution and traffic congestion).

In Taipei, the urban regeneration and development outcomes are similar to Reading. The development orientation of Taipei originally was planned to develop from the west, the old centre, and then extend to the east, the new centre, but has been modified to focus back to the western side of the city, mostly the older and more deprived areas. Yet the strategy of using brownfield land and maintaining the area's cultural heritage has successfully revitalized the western area. It seemed to be a win-win-win initiative for the environment, the economy and society. The urban regeneration policy meets the sustainable eco-city concept with futurity and eco-system principles by using brownfield development. However, it has been a predominantly government-led regeneration without much community involvement. Some of the NGO members argued that *"the urban regeneration did not consider the citizen's opinions. When moving citizens from old areas to new areas, TCG did not allow citizen participation. They ruined the community spirit in the process of urban regeneration. Moreover, some of the pavements were roughly and unattractively constructed with concrete without concern for design or the environment. Rain water could not penetrate them at all".* It is evident that urban regeneration in Taipei City has faced some significant problems in terms of citizen participation and environmental issues.

Urban regeneration in both case study areas seem to have "successfully" achieved economic development and created some elements of a win-win-win initiative for the environment, the economy and society where the public sector, private business and the community have all benefited. However, both focus more heavily on the economy than on the environment. Taipei further suffers from very limited community involvement.

4. Green Building Assessment has been Used to Promote Green Design, but it Is still Under-developed

In order to help achieve sustainable eco-city objectives, green building assessments have been developed by the respective governments to promote green design in both their areas. Both assessments consider health, recycling and ecology, and have had some success.

In Reading, the EcoHomes standard was introduced by the Building Research Establishment (BRE) in 2000. EcoHomes contributes to the achievement of sustainable development aiming to provide a better quality homes for now and future generations. All schemes receiving funding from the Housing Corporation's National Affordable Housing Programme are required to achieve an EcoHomes 'Very Good' rating from April 2006. EocHomes covers a range of environmental issues which include (Wilson and Smith, 2006, p. 6):

- *Energy*
- *Transport*
- *Pollution*
- *Materials*
- *Water*
- *Land use and ecology*
- *Health and well being.*

It is a very important housing standard which can be used by Reading Council to help achieve their eco-city objectives. Reading's affordable housing has so far only been graded mostly 'Good'. One council officer said, *"We try and obtain Eco Homes 'Very Good' on schemes if possible but in fairness, mostly we obtain 'Good'"*. Perhaps Reading needs to put more effort into this initiative with more attention paid towards the Eco-Homes grading of private housing. EcoHomes consider a range of environmental issues which can help reduce resources consumption, increase resource efficiency and help Reading achieve sustainable eco-city objectives.

In Taipei, there are two assessment systems developed to evaluate buildings, both of which are administered by central government, and are currently top priority for the government in its push towards the implementation of its sustainable development policy.

The first assessment system was the Green Building Evaluation and Labelling System which began in 1999 and was established to accredit new private and public buildings using nine indicators (i.e. site planting, water resource use, site water preservation, daily energy saving, CO_2 reduction, waste reduction, and sewage and rubbish improvement). Up until now there have been 815 certificates issued, with 9,522,140 sq. m. of floor space now coming under the Green Building Label throughout Taiwan. Government predicts that the building energy consumption has been reduced by 20% and water consumption reduced by 30% in these buildings (Chinese Architecture and Building Centre, 2005).

The second achievement was developing a Green Building Material Labelling system which started for private enterprises and factories in July 2004. It targeted four aspects: health, recycling, high performance and ecology. Two companies have been successfully awarded the Green Building Material Label so far.

It is apparent that the green building schemes are useful aids on the path toward a sustainable eco-city. It is suggested that not only the public sector but also the private sector could put more effort into both schemes, and with more government encouragement. However, it has to be said that these initiatives are rather 'technocratic' in approach and fail to fully consider that construction is very much a social process involving different actors and interests (Rohracher, 2005, p. 202).

5. Green Transport Is Potentially One of the Most Useful Tools, but it Is not very Successful in these Two Contexts

It is evident in both cases that green transport has been recognised as one of the main tools with which to achieve the sustainable eco-city. Both studies have made some efforts in policy-making and implementation in areas such as public transport; yet, because of certain constraining factors, particularly local politics and national cultural attitudes to car ownership and use, the implementation has not been as effective as intended.

Table 4c. Green Transport Outcomes: Comparison Between Reading and Taipei

Policies	Reading	Taipei
Sustainable Transport and Subsidized Policy	demand management approach non-subsidised policy on train and bus tickets and petrol prices Nearly the highest petrol prices in the world Proportion of travel to work by bus ↓ from 15.8% in 1991 to 12.2% in 2001	Central control subsidised Subsidised train and bus tickets and petroleum price The cheapest petroleum price among the four economic dragons in Asia. Number of motorcycles ↑ Highest motorcycle numbers in the world 1,026,726 motorcycles
Green Travel Plan	demand management approach Work with local businesses and organisations to develop and implement Travel Plans to encourage trips to work and for business by public transport, walking and cycling	None
Car Parking	demand management approach Parking pricing, reducing parking spaces and Park and Ride Parking space for cars are decreasing every year Peak time traffic congestion	Encourage the public and private sector construct parking lots Low parking fee Parking space for cars and motorcycles are increasing every year Peak time traffic congestion
Pedestrianisation	Most of the shopping area in the town centre is pedestrianised. The length of pedestrianised street total: 1,046 metres	the policy of Taipei claimed "Pedestrian-Oriented Designs - Ensuring Pedestrian Safety" TCG still endorsed the vehicles and tried to meet their needs. Taipei's street furniture disorderly and dangerously upright People have no concept of pedestrian rights
Cycling	Total length of cycle routes in Reading is 78.8 km, of which: traffic Free - 42.1km, On Road - 5.795km, Signed only - 31.0km. in addition, 33.9 km of additional length of cycle route is planned to be completed by 2010. The proportion of travel to work by bicycle has slightly increased from 2.8% in 1991 to 4.1% in 2001.	It is very dangerous to cycle on the road mixed with tremendous number of high speed cars and motorcycles. Most cyclists ride on the pavements, but the government declared in Sep. 2005 there would be fines for cycling on the pavements from 2006.

For example, RBC has promoted the local bus company, Reading Buses, to encourage more bus use. Reading Buses have made some significant performance improvements, such as: major investment in new, more comfortable and more environmentally-friendly vehicles; real time information provision at bus stops; and more high quality shelters. Moreover, Reading has helped produce a number of creative green transport plans aimed at a demand management approach to reduce resource consumption. On the other hand, Reading's efforts to reduce car-use has shown little progress due to residents' strong car-dependency and lack of any really effective public transport subsidy (the latter reflecting Central Government policy).

Taipei has facilitated some significant public transport schemes such as the Mass Rapid Transit system (MRT), and a convenient bus system. It introduced the Taipei Easy Card system in 2002 and integrated MRT, city buses and public parking lot payments. It successfully encouraged greater citizen use of the MRT, but bus use has declined. However, at the same time the TCG has also improved road infrastructure and adopted a subsidised petroleum policy which has allowed a great number of motorcycles in the city. Although all TCG and councillors recognised the problems of increased traffic, they felt they could not solve the problems without concern for the impact on political votes. Unsurprisingly it has led Taipei towards being a vehicle-centred city, with limited regard for pedestrian rights.

The main policy and implementation outcomes in the two case studies are compared in Table 4c. It is apparent that both cities have attempted to reduce travel and private car use by a number of environmentally sound policies and initiatives. Yet some of these initiatives (such as subsidised petroleum, motorcycle policy, and liberal car parking policies) have resulted in the transport problem getting worse. The culture of car-dependency and local political sensitivities are also identified as two of the main factors preventing effective policy implementation and, consequently, delaying the pace of local movement towards the sustainable eco-city. The few positive outcomes in this area are relatively insignificant.

6. Waste Management and Recycling Is also a Significant Tool, although Some Issues Exist

It is certain that waste management and recycling is one of the main tools that can be used to further sustainable eco-city objectives. The different local contexts and cultural perceptions in the UK and Taiwan have also played their part in producing a variety of outcomes.

Until quite recently Reading has not paid much attention to waste and recycling issues. However, since 2003 the council has worked co-operatively with Bracknell Forest Council and Wokingham District Council through the re³ partnership to develop shared facilities that will manage waste for the next 25 years. Landfill is the most urgent problem, as the one remaining site is likely to be filled by 2008. However, it is not easy to find new landfill sites due to increased public environmental awareness and limited land resources. Although Reading's recycling rate has significantly increased to 27% in 2006/2007, 70% of the waste (over 47,000 tonnes) is still disposed of in holes in the ground – as landfill (Reading Borough Council, 2007).

Taipei has tried to tackle the waste management and recycling issues for some years now. They introduced the categorised recycled rubbish programme in the 1990s and a stricter approach from 2000 involved every household in this programme together with a rubbish collection charge on a per bag basis. Significantly the programmes were very successful. In addition, the household kitchen waste recycling programme has been enforced since December 2004 leading to further reductions in waste. In addition, since July 2005, every shop in Taipei is forbidden to provide free plastic bags for customers, or risk being fined. It is evident that Taipei City has tried to use demand management to reduce the consumption of resources. Somewhat surprisingly, local citizens have followed the rules and the programme has reduced the amount of waste produced each year by a total of 53%.

The analysis of the empirical study outcomes in both cases shows most of policies and visions cover some of the four principles of the sustainable eco-city model. However, in practice some of the anticipated outcomes have not yet been reached. The different outcomes depend on policy making and implementation in different local contexts which are affected by key factors. Generally they all use land use and transport as the main tools in the drive towards a sustainable eco-city and normally a demand management approach is adopted. Taipei has paid more attention to environmental protection and waste recycling and management than Reading and has focussed on a technology-based approach and a vehicle-centred set of policies, rather than a people-centred, participatory, approach. Reading, on the other hand, strongly influenced by New Labour's Third Way, has put more emphasis on participation and equality than Taipei. Reading's strong emphasis on economic competitiveness has, directly or indirectly, affected the policy making and implementation processes. Significant attention has been paid to developing a 'sustainable' economy and, to a lesser extent, sustainable transport, and has relied upon a politician-led strategy though with a relatively long tradition of participatory democracy.

In conclusion, both case studies have illustrated a number of significant eco-city policies, but policy implementation has not reached full fruition yet as they have been hindered by many socio-political, cultural and economic factors operating at local and higher levels.

SUMMARY

It has discussed the results of the two empirical studies as a whole with reference to the theoretical propositions of the sustainable eco-city framework. As a result, it has been found that the two empirical studies provide evidence to demonstrate that planning for the sustainable eco-city can be analysed in terms of the proposed new framework.

In the first section, the empirical evidence has been reviewed and discussed how this sustainable eco-city model has been defined and interpreted at a local city level. The discussion explored the context and perception of sustainable eco-city ideas to understand how they can be affected by political, economic, institutional, geographical and cultural contexts, both locally and globally. The context and perception of sustainable eco-city ideas have been explored in the two empirical studies.

Two main findings have been reported. First, the empirical evidence has shown that local and global contexts can have a strong influence on the perception and policy evolution of a sustainable eco-city. These contexts include political, economic, institutional, geographical and cultural factors which could also be identified as influential in affecting the sustainable eco-city. For example, these factors in the local and global contexts are intertwined and interdependant and influence the relationship between local and central governments as they seek to co-operate in the planning process.

Secondly, the perception of a sustainable eco-city was interpreted depending on the practitioners' knowledge and involvement and once again, was affected by the local and global contexts. The analysis of the two empirical studies provides evidence to support this ontological assumption. The nature of sustainable eco-city policy is constructed by individuals engaged in local practice, such as the academic in Reading who has long been involved in local policy making since drafting Reading LA21.

In the second section, the empirical evidence has been discussed to understanding the planning process of the sustainable eco-city. It was found that planning can be explained successfully as a process of networking and consensus-building between different stakeholders. Some findings from the empirical evidence were discussed around the following issues:

First, during the planning process, the following trends were identified in the empirical evidence:

- formulation and evolution of eco-city policy has gradually embraced the four principles;
- some key influential factors are identified during the sustainable eco-city policy evolution, such as political, economic, institutional, geographical and cultural factors;
- planning systems have gradually been revised towards sustainable eco-city;
- both case studies show movement towards decentralised decision making, but both central and local governments are afraid that this decentralisation will result in a reduction of to themselves; and
- There has been a movement from local government to local governance. The transformation of local government to local governance reflects the intention of central government to decentralise responsibility and empower local government and local residents, albeit within a framework of controls and targets.

Secondly, some key mechanisms in the planning process have been identified as follows:

- community participation;
- community network and consensus-building;
- the dynamics of power and influence;
- participatory and representative democracy;
- the adoption and significance of the Third Way approach;
- training and capacity-building for the community; and
- the equality principle.

In the last section, the empirical evidence has been discussed in the light of the outcomes of the eco-city planning processes. A number of conclusions were reached:

First, it is evident that the analysis of the outcomes in both cases shows most of the policies and visions cover some of the four principles of the sustainable eco-city model. However, in practice some of the outcomes have not effectively achieved their aims yet as they are affected by many components including local socio-political, cultural and economic factors. These will be examined in more depth in the next chapter.

Secondly, the different outcomes depend on policy making and implementation in the different local contexts which are affected by some influential factors. Therefore, the policies and visions cover some of the four principles, but to different extents.

Thirdly, they generally both utilise land use and transport as the main tools for progressing sustainable eco-city policies and normally the demand management approach is adopted.

Fourth, usually the Sustainable Development Indicators are used to evaluate the performance of the sustainable eco-city. The local level indicators appear to better reflect the principles of the sustainable eco-city. Also, some nationally-derived indicators in Reading are designed and used to evaluate local government performance with feedback influencing central government grants and other spending programmes. It is apparent that these indicators reflect the different values of various interest groups and should therefore be treated with some caution.

The comparisons of two projects using the empirical evidence offers a new perspective to understanding the planning processes and outcomes of a sustainable eco-city and provides an understanding of how to improve things in practice and resolve its problems.

REFERENCES

Breheny, M., and Hart, D. (1994). Reading in Profile: A Survey of Key Economic Issues for the Greater Reading Area in the 1990s. Reading, Reading Borough Council

Chinese Architecture and Building Centre (2005). Green Building Label. Chinese Architecture and Building Centre

Doak, J. (1998). Changing the World Through Participative Action: The Dynamics and Potential of Local Agenda 21. IN COENEN, F. H. J. M., HUITEMA, D. & LAURENCE J. O'TOOLE, J. (Eds.) *Participation and the Quality of Environmental Decision Making.* Dordrecht/Boston/London, Kluwer Academic Publishers

Driver, S. and Martell, L. (2000). Left, Right and the third Way. *Policy & Politics,* 28, 147-61

HM Treasury (2005). Thames Valley Economic Partnership - Thames Valley: Sustaining Our Success. HM Treasury

Hsiao, H. (2005). Yin-Jiu Ma's " Third Way". *China Times.* Taipei

Raco, M. (2003). Remaking Place and Securitising Space: Urban Regeneration and the Strategies, Tactics and Practices of Policing in the UK. *Urban Studies,* 40, 1860-1887

Reading Borough Council (2003). Review of Reading Borough Local Plan (2001-2016). Reading, the UK, Reading Borough Council

Reading Borough Council (2006). Reading Central Area Action Plan Consultation on Issues and Options. Reading Borough Council

Reading Borough Council(2007). Reading Waste and Recycle. Reading Borough Council

Rohracher, H. (2005). Social Research on Energy-Efficient Building Technologies Towards a Sociotechnical Integration. IN GUY, S. & MOORE, S. A. (Eds.) *Sustainable Architectures- Cultures and Natures in Europe and North America*

Sun, S. (2008). Planning for a Sustainable Eco-city, PhD thesis in Department of Real Estate and Planning, the University of Reading Business School, England, UK

Wilson, C. and Smith, B. (2006). *EcoHome-Achieving Very Good,* Teddington, Middlesex, Housing Corporation

In: Eco-City and Green Community
Editor: Zhenghong Tang

ISBN: 978-1-60876-811-0
© 2010 Nova Science Publishers, Inc.

Chapter 5

PRESERVATION OR DEVELOPMENT – TDR AND THE APPLICABILITY TO URBAN GREENBELT OF SEOUL

Yunwoo Nam

Community and Regional Planning, University of Nebraska,
Lincoln, NE, USA

INTRODUCTION

Over the world, many cities are often faced with the challenge of trying to encourage the development and expansion of built-up areas, while, simultaneously protecting the historical resources and natural environment. The conflict between preservation and transition is a critical issue when considering eco-cities and sustainable development. This chapter responds to the growing interests in the encouragement of urban growth strategies that are sustainable environmentally and socially. Two programs, TDR and Greenbelt, are introduced and discussed.

One of innovative smart growth management tools used to reduce the conflicts between these seemingly incompatible objectives is called 'Transfer of Development Rights (TDR)'. This technique allows development rights, which are unused on one parcel of land to be separated from that parcel and transferred to another parcel. The TDR program can be used to achieve numerous land use goals ranging from growth management to the preservation of open space, historic landmarks, natural areas, and agricultural land. In American planning practice, a TDR program has been a useful tool to reduce the financial inequalities that stem from land use regulation and a compromise means of compensation to the landowners.

Another tool of urban growth management is known as 'Greenbelt' policy. The greenbelt is an urban containment strategy to designate areas of open space surrounding the rapidly growing city. It is designed as a growth management program against urban sprawl to direct the most development into existing urban areas inside the boundary and to preserve ecosystem. In an urban environment, a green space is crucial for the well-being of residents due to its diverse ecological functions (Fung and Conway 2007), such as air pollution filtration and the biodiversity conservation.

In Korea, the greenbelt system was introduced in 1971, and has been a main target of arguments over environmental protection and urban development. The designation of the boundary lines was decided on political reasons, rather than as an outcome of environmental impact assessment or land use surveys. It also has compensation problems for the loss of individual property rights regulated by the government for the purpose of public interests.

The TDR principle may be applicable to other countries' situations, especially Seoul's greenbelt case. In principle, TDR creates a market oriented mechanism as an alternative to regulatory approach of zoning. It provides incentives for compliance. It motivates the conservation on private lands by compensating land owner's opportunity costs of sacrificing development. This chapter reviews the theory and practice of TDR, discusses the key factors of a successful program, and explores the potential for its use to conserve natural areas in Seoul metropolitan area.

This chapter mainly consists of two parts: introducing TDR as an innovative growth management program and the applicability of TDR approaches for Seoul's greenbelt case. In the first part, I review the American experiences of TDR from a theoretical perspective as well as a managerial perspective. In the second part, I review Seoul's greenbelt policy, discuss unexpected effects, and explore the potential of TDR applicability for environmental planning and open space preservation.

TRANSFER OF DEVELOPMENT RIGHTS: A SMART TOOL OF GROWTH MANAGEMENT

Concept of TDR

Transfer of Development Rights is a variation of traditional zoning practice. TDR is often used when it is desirable to limit development in a particular area to protect cultural, agricultural or natural resources, such as historic landmarks (Henderson 1998), farmlands (Peters 1990), wetlands, open spaces, or forests (Chomitz 2004), and to concentrate development in another area. Specifically, instead of simply imposing limits in sensitive areas, this approach recognizes development rights in the areas where development is limited but allows the rights to be used in other areas. The basic process is initiated when a municipality designates an area of open space and prohibits development therein. Then, residential development potential in that area is transferred to another area where the residents agree that development is feasible. Land owners in the preserved areas, who will continue to own their land, may sell their rights for further development to other landowners, developers, or builders who like to develop areas proposed for development.

The concept of TDR is based on the "bundle of rights" theory of property ownership. Under this theory, landowners possess a bundle of rights including the right to use, the right to dispose, and the right to develop on the property (Pizor 1978). Each of which can be separated from the raw property and be transferred to another as a commodity (Ziegler 1995). However the ability of private landowners to exercise their rights, particularly their development rights has been constrained by the police powers, the power of eminent domain, and the common law prohibition on conducting a public or a private nuisance (Pedowitz

1984). In the landmark case of *Village of Euclid v. Ambler Realty Company*,[1] the police power was expanded to encompass various land use controls and many governmental limitations on the property began to be imposed in the name of zoning (Levy 2008). The enactment of zoning law under police power limited the development potential of the property to only certain uses and in the way thereby permitted. So, there were conflicts between zoning and individual profits, because land use planning as implemented by police power regulation inevitably challenges the profit motives of individual owners, which are soundly supported by constitutional protection. Constitutional prohibitions on the 'taking' of property without due process of law required zoning designations and other land use controls to pay just compensation (Sax 1971, 149). As the number of taking cases increases, the financial burdens of eminent domain began to bear more heavily on municipalities who sought to preserve landmarks, environmentally sensitive sites, and farmlands. This background gave rise to TDR as an innovative concept.

TDR attempts to preserve historically and environmentally important areas without violating basic rights and due process under the constitution. It also attempts to avoid the taking problems by providing compensation for property owners through the sale of development rights. It combines planning with certain aspects of property law (Chavooshian and Norman 1973, 12-13).

There are many variants of the TDR concept to achieve specific goals. For example, in the New York and Chicago plan, TDR was used for the preservation of architectural and historical landmarks by transferring the right to develop that land more intensively to owners of other land (Woodbury 1975). In the New Jersey program, it was used to preserve farmland and open space by transferring the right to develop that land to designated districts (Gottesegen 1992; Pizor 1986). In Collier County program, Florida, the TDR program is designed to preserve open spaces and ecologically sensitive areas – specifically islands, marshes, and coastal areas outside of Naples (Pruetz 1996).

Major Components of the TDR Program

Even though there are various TDR programs in U.S, they are not the same programs. However, they have common essential factors.

1. Sending Area

The area from which the development rights are transferred, and is sought to be protected from development is called the sending area. The sending area can vary with the purpose of the program, and may involve farmland, forested areas, historic sites, recreational sites, wetlands, aquifers, coastal areas, scenic landscapes, etc. When a sending area is designated, the zoning authority will reduce the degree of development to be permitted within that area by amending the zoning ordinance. Many successful TDR programs strongly encourage transfers by making development difficult or impossible on sending sites. By controlling the amount of allowable development, the sending area is protected from the undesirable effects of development and, at the same time, that development is encouraged to prosper elsewhere in the region.

---

[1] *Village of Euclid v. Ambler Realty Company,* 272 U.S. 365, (1926).

The potential sending area may or may not have already contained some degree of development. The degree of existing development in the sending area is an important issue because it influences the expectations of property owners with regard to the market value. Developable lands in compact, dense and attractive urban settings may be higher-priced than similar lands in the sparsely populated countryside. This matter regarding a taking issue when downzoning in the sending area reduces the potential value of the property significantly (Stinson and Murphy 1998).

2. Receiving area

The receiving area accommodates development potential from the preservation (sending) area. In adopting a TDR program, a community designates as receiving sites areas of land where the increase in density resulting from the transfer would be beneficial and acceptable to the community. Receiving areas must be subject to sufficient growth pressures, and development regulations in these areas must be restrictive enough to provide landowners with an incentive to purchase development rights.

Successful programs often provide a strong incentive to transfer development rights by making it difficult or impossible to achieve additional density on receiving sites without using the TDR method (Pruetz 1997). Before receiving and sending areas are designated, an environment impact analysis must be conducted.

3. Transferability and Marketability of Development Rights

In most TDR programs, the development potential, mostly known as 'rights' or 'credits,' may be detached from the property and marketed separately from the land. The credit is the feature of a TDR program, because it is the element that mitigates the impact of preservation zoning in the sending area. It provides compensation mechanism to balance the gains in land value that accrues to landowners in the designated growth areas against the corresponding financial loss in value experienced by landowners in the preservation areas. In this way, TDR incorporates a sense of fairness into traditional zoning process (Gottsegen 1992, 10). The method of allocating development rights (credits) and regulating their use are policy decisions made at the municipal level and thus, may vary considerably with each TDR program.

To be a successful TDR program, the crucial element is the marketability for the trading of TDR rights. It should be ensured that a TDR market exists and that TDR have value so that there is sufficient incentive for its use. If the market is not active, TDR rights may not be sold. Without trading, the TDR program cannot accomplish its goal. Marketability is ultimately expressed in terms of the dollar value of the rights. If the price of the right is too high, the developers will not use them. In a more active market, owners of TDR rights are likely to receive a reasonable price. As rights increase in value, sellers will be eager to participate in the program, because they perceive the value of their rights to be matching with the opportunity cost by not developing their sites. When the value of rights falls short of expectations, property owners in the sending area may complain of unfairness. Therefore, the biggest challenge to implementing a successful TDR program is to facilitate participation and encourage active trading in the TDR market. However the value in real estate properties is an outcome of complicated interactions of many factors. For example, the location and community characteristics often influence the different values of similar properties.

4. Legal Framework

The TDR program needs the legal framework that defines the receiving and sending areas, describes the credit allocation procedure and establishes the procedure for the transfer of development potential from the sending area to the receiving area. This framework should be compatible with the municipality's comprehensive plan and zoning ordinance (Gottesegen 1992). In communities that plan to use TDR as the basis for channeling all future growth to compact development centers and to preserve valued local resources, a revised comprehensive plan should be prepared that incorporates the fundamental aspects of the proposed TDR program. The comprehensive plan should provide the underlying basis for the use of TDR and its relationship to the community's comprehensive land use goals (Gottesegen 1992; Machemer and Kaplowitz 2002; Tainter 2001).

5. TDR Bank

A TDR bank is a public or quasi-public institution[2], which buys development rights and then sells them to the developers of receiving areas. While TDR banks are not required for well-structured TDR programs, it can serve as an important support for landowners and developers (Johnston and Madison 1997; Gottesegen 1992).

With the TDR bank, sending site owners can sell their TDRs at any time rather than waiting until those TDRs are needed for a receiving site project. On the other side, receiving site developers can purchase them from the TDR bank rather than trying to persuade a sending site owner to sell TDRs (Pruetz 1997). In addition to connecting buyers and sellers, TDR bank can increase the certainty and the speed of transfers which are important for the decision of developers (Machemer and Kaplowitz 2002, 789). TDR banks can influence the prices by controlling their selling and buying. This means that a TDR bank can serve a valuable role in moderating extreme market conditions that may arise in a TDR program and alleviating hardship conditions for landowners.

TDR Planning Network

TDR involves all the participants of the conventional land use planning process – municipal, county and state officials, landowners, farmers, developers and the general public. Each group of participants brings its own goals and needs to the planning network and works to have those interests met. This planning process is where many of the politics of growth management are found.

The stakeholders in the TDR process are various (Pruetz 1996; Sungu 1996; Taintor 2001). The general public's interest is based on a desire for an increased quality of life resulting from TDR (e.g. better neighborhood, good air quality, safe water supplies, and open spaces etc.). Landowners are another group, whose properties are targeted for preservation or development. Since landowners' primary interest tends to be equity and the ability to liquidate it, they take a very active role in the planning process. Farmers may want to ensure that high quality agricultural land is obtainable at an affordable price. Developers are also

---

[2] For example, New Jersey Pinelands case created two government agencies to facilitate TDR transfers. The sate legislature created the Pinelands Development Credit Bank, and local government created the Burlington County Pinelands Development Credit Exchange.

stakeholders because they try to realize the highest net profit possible from the new development. Homeowners in the community planning for TDR are interested in having their community maintain its high quality through attractive neighborhoods, well-maintained parks and open spaces. Local city governments and planning boards are charged with meeting the combined interests of the residents. County and state governments try to balance the development interest with the broader public goals such as environmental protection and natural resource preservation. All these stakeholders interact on the planning network.

Because of many actors and their interests in the planning process, a TDR program process looks like a political compromise between development and preservation. Each TDR program will fall somewhere on a continuum between complete development of the community and complete preservation. Exactly where on that continuum an adopted program falls depends on the goals it is designed to achieve and the role of the stakeholders involved in the TDR planning network. A well-designed TDR program can meet the majority of the interests of all the participant groups.

Functions of TDR

1. Balancing Development with Preservation

TDR is a land-use tool that addresses one of the key issues of growing urban areas: namely, how to accommodate pressures for growth and development and at the same time preserve essential resources such as water supplies, farmland, and important features such as landmarks. The problem can be stated clearly. Land near the big cities is being converted to urban uses at a rapid rate. The result is a wasteful consumption of the nation's limited land resources.

The dilemma that confronts land-use planners is to devise a way of balancing urban growth with environmental preservation. TDR has been advocated as a resolution to this dilemma. TDR programs may guide a systematic approach to help communities manage comprehensive and long range environmental and economic development goals. The use of TDR permits the designation of broad areas as preservation areas, which may include those natural resources that are important to the community. TDR provides a realistic tool that can be used for effect preservation at a lower cost than other tools available to local governments.

2. Mitigating Windfalls / Wipeouts Dilemma

A windfall/wipeout is "an increase/decrease in the value of real estate not caused by the owner or by general inflation/deflation" (Barrese 1983, 235). If a wipeout occurs due to a governmental regulation, the loser may charge that the government has unjustly taken his property without compensation.

The implementation of land use regulations on privately owned lands often generates imbalance of unjustified gains and losses in the private sector. While the preservation of landmarks, environmentally sensitive lands, and farmlands cause wipeout threats, the designation of high-density developable area gives windfall fortunes to the landowners. TDR breaks windfalls/wipeouts dilemma (Gans 1975, 275-276). Under TDR programs, owners of restricted resources are not wipe-out, but are properly compensated by selling development rights and the windfall of increased land values that land owners in transfer areas might otherwise enjoy in consequence of these restrictions in offset by the payment they make for additional development rights (Costonis 1973, 99).

3. An Innovative Growth Management Tool without Public Expenditure

Often compensation may be needed to adopt land use controls that actually achieve a community's goals. One method of compensation is to acquire the property using grants or other public funds. However this method is difficult to achieve goals completely because of limited budgets.

Transfer of development rights helps a community plan effective zoning. The net effect of TDR is the preservation of environmentally important areas with equitable compensation for owners. However, there is no cost to the taxpayers since no acquisition by government is involved. This compensation factor allows communities to adopt more effective zoning requirements to help communities meet numerous land use goals from growth management to the preservation of open space, historic landmarks, natural areas, and agricultural land (Pruetz 1997). And, at the same time, it also can function to channel development forces to predetermined locations where extensive capital improvement programs are proposed or underway, thereby serves as a catalyst to improve land use planning (Kwon 1996). By designating sending areas and receiving areas, market demands for new constructions can be guided to those areas in a comprehensive land use scheme.

Issues of TDR Programs

1. Administrative and Political Issues

The TDR proposal is based upon the assumptions that the owners of preserved open space land will be compensated for the deprivation of the use of their land by the sale of certificates of development rights to owners of developable land. This relationship is predictable in theory and workable in practice only to the extent that the planners' projection of future economic demand for land development is accurate and their designation of sites for specific land use is skillfully performed. It means that the success of the program may well depend upon the planners' proficiency in this task.

Even in the case that the planner has performed his job well and submits his findings and recommendations to the government body, there will be political pressures imposed upon the government body of elected officials to modify the original recommendations to enhance political as well as planning significance (Rose 1975).

2. Legal Issues

The primary legal problem associated with TDR implementation is that the courts find that a TDR program violates a section of the fifth[3] and fourteenth[4] Amendments of the United States Constitution known as the "taking clause".

---

[3] U.S.Const. Amend. V.: "No person shall .... be deprived of life, liberty, or property, without due process of law; nor shall private property be taken for public use, without just compensation."

[4] U.S.Const. Amend XIV. Section 1: "... No state shall make or enforce any law which shall abridge the privilege or immunities of citizens of the United States; nor shall any State deprive any person of life, liberty, or property, without due process of law; nor deny to any person within its jurisdiction the equal protection of the laws."

The local government's authority to regulate the use of private property through the adoption of zoning measures derives from a state's police power to protect the public health, safety, and general welfare (Levy 2009, 75). However, despite its broad authority to regulate pursuant to these police powers, government action remains subject to the fifth and fourteenth Amendments of the United States Constitution, which preclude government from taking private property for public use without just compensation, and from depriving individuals of their private property without due process of law. Thus "while property may be regulated to a certain extent, if the regulation goes too far, it should be recognized as a taking."[5]

The TDR concept was granted judicial approval in the 1978 Landmark case of Penn Central Transportation Co. v. City of New York.[6] In finding that the Landmarks Law did not amount to a taking of Penn Central's property, the Supreme court held that the regulation would not interfere with its air rights because its ability to use these rights was made transferable pursuant to the TDR ordinance. The Court stated ".... While these (transferable development) rights may well not have constituted "just compensation" if a "taking" had occurred, the rights nevertheless undoubtedly mitigate whatever financial burdens the law has imposed on appellants and, for that reason, are to be taken into account in considering the impact of the regulation...." (Pruets 1996, 17). This decision suggests that TDR may be used to address potential impacts of restrictive zoning regulations as a mitigation mechanism which should be considered in determining whether there has been a deprivation of all economically beneficial use of property or undue interference with the property owner's reasonable investment-backed expectation.

According to the Penn Central decision and other recent regulatory taking cases, the adequacy of the TDR program as mitigation or compensation depends in large part on how valuable the transferred development rights are perceived to be. Therefore, there is a need for a well-planned system with marketable TDR which would reduce the risk that a taking lawsuit will be filed.

3. Economic Issues

The marketability of TDR can increase the ability of its program to mitigate the economic impact of the restrictions effectively. In principle, TDR fully compensates the owner of the restricted lands for whatever development rights he may lose as a result of landmark or open space designation. However, in practice, there are many economic factors to be considered for the success of TDR. This mitigation mechanism can be accomplished in the active TDR market. For the vital and healthy market, first, the price of development rights should be attractive enough to approach to full compensation. Second, the development on the sending site should be highly restricted. It will cause the increase of TDR supply. In addition, the receiving site should also include both the demands for additional developments and the land use controls which make that additional development possible only or primarily, through TDR. It will lead to the increase of TDR demand.

---

[5] Pennsylvania Coal Co. V. Mahon, 260 U.S. 393 (1922).
[6] Penn Central Transportation Co. V. City of New York 438 U.S.104 (1978).

SEOUL'S GREENBELT POLICY

Most growth management programs of local governments aim to accommodate development while maintaining the quality of life in the community and preserving environmental qualities. Following London's successful experience in containing the expansion of urban built-up areas, preserving agricultural lands, and protecting the natural environment, the greenbelt program has been widely applied in various cities and countries (Amati and Yokohari 2007; Tang, Wong and Lee 2007). The greenbelt is a popular land use planning program as a smart growth management tool. But the greenbelt program also brought the unexpected outcomes of social and economic costs such as the shortage of urban developable lands, the higher density of the inner boundary, the increase of land and housing prices, and the unfair restriction of private property use. This section examines Seoul's greenbelt policy.

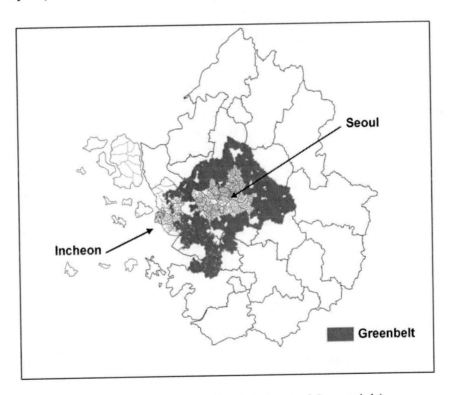

Figure 5a. Greenbelt in Seoul Metropolitan Area (Seoul, Incheon and Gyeonggi-do).

Review of Greenbelt History

1. Rapid Growth of Seoul

The Seoul Metropolitan Area (SMA) is the most dynamic and rapidly growing urban agglomerations in the world. It has a modest origin as a walled capital of the Yi Dynasty (1392-1910). For the past 600 years, the city has remained dormant and laggard without undergoing significant change.

Table 5a. Population Growth and Concentration
(unit: 1,000 persons, %)

	1960	1970	1980	1990	2000	2005
Nation	24989	30852	37407	43520	46136	47278
Seoul	2445	5536	8364	10627	9895	9820
	(9.8)	(17.9)	(22.4)	(24.4)	(21.4)	(20.8)
SMA[1)]	5198	8731	13279	18583	21354	22766
	(20.8)	(28.3)	(35.5)	(42.7)	(46.3)	(48.2)

Source: EPB, 1960, 1970, 1980, 1990, 2000 *Population and Housing Census Report*.
NSO. 2007. *Social Indicators in Korea*, Seoul.
Note: 1) SMA denotes Seoul metropolitan area and includes Seoul, Incheon and Gyeonggi-do.

However, since the early 1960s, in parallel with Korea's drive for industrialization, Seoul began to acquire enormous momentum for growth and structural transformation. Industrialization triggered the impetus for nationwide geographical movement of people from rural to urban areas. Traditionally, most economic, political, educational[7] and cultural bases have been located either in Seoul or in the surrounding Seoul Metropolitan Area, thereby encouraging more people to migrate to the capital.

The urban population growth of Korea, as illustrated in Table 5a, can be characterized by its speed and unique pattern. Not only has urbanization proceeded very rapidly, but it also occurred in a concentrated fashion in a small number of large cities.

In 1960, the urban population in Korea was 6,997,000 persons that were about 28% of the population in the nation as a whole (Kim and Choe 1997, 190). Since then, urban population has grown consistently and rapidly. It amounted to 32,397,000 in 1990 that account for 75% of the national population (1990 Population and Housing Census). The tremendous growth of the urban population, 360% over 30 years (1960-1990), is undoubtedly regarded as extremely rapid, compared with the experiences of other countries. Most conspicuous in this growth trend is the concentration of population in the Seoul metropolitan area. In 1960, the population of the city of Seoul was 2,445,000 amounting to just 9.8% of the national population and became 10,627,000 in 1990 (24.4%), and 9,820,000 in 2005 (20.8%). If we consider the population of the Seould Metropolitan Area (SMA), the concentration of population becomes more prominent. While the population in the SMA accounted for 20.8% of nation's population in 1960, it rose to 42.7% in 1990, and 48.2% in 2005. Thus, almost half of the national population was residing in SMA. Since the early 1990s, the population of Seoul has decreased a little. However the total population in the SMA increased, as the populations of neighboring suburbs increased. In 2005, Seoul had 20.8% of the national population. However the population of SMA reached 48.2%. The concentration of population can be more clearly revealed by the statistics on population density. The population density of the nation as a whole was 476 persons/km^2 in 2005. That of Seoul was 16222 persons/ km^2, 34 times higher than the national average. Along with this accelerated rate of increase in

---

[7] The educational system in Korea is heavily centralized. In general, the most prestigious universities are still located in Seoul, and most people seem to believe that the quality of education for primary and secondary schools is better in Seoul. See M.Hwang, Growth and management of Seoul Metropolitan Area, presented to conference on Urbanization and National Development, Honolulu, HI, (January, 1990).

population, the physical size of Seoul has also been growing rapidly. Consequently, Seoul has encountered severe problems of housing shortage, inadequate physical infrastructure and thus, needed growth control measures.

1. Greenbelt Designation

To alleviate the socio-economic problems due to the excessive concentration in Seoul, policy makers had tried to control the further expansion of the Capital region with a designation of greenbelts, where virtually no development is allowed around the major urban centers. The greenbelt, which was formally named as the development restrict zones,[8] was initially designated in 1971 (City Planning Act of 1971),[9] and expanded until 1976, and have been maintained with little changes in their physical configuration and policy, until 1990s. Although the concept of the greenbelt was borrowed from London's greenbelt case, planners transformed the practical principles to suit local circumstances. Since the 2000s, the greenbelt policy was revised, and parts of greenbelts have been released. The central government and local governments are trying to convert the land use in the greenbelt region to urban uses for various reasons. When the greenbelt was designated, the primary purpose was to control the chaotic urban sprawl, to secure healthy living environments surrounding urban fringe areas, and to limit urban development for some national security reasons if necessary. However, due to environmental aggravation along with the rapid urban growth, the focus of greenbelt policy is shifting from urban growth control (anti-urban sprawl) to environmental conservation programs including the natural environmental preservation and the prevention of environmental degradation.

The total greenbelt area is 5397 km^2 nationwide, about 5.4% of Korea's total land area (Kwon 2004, 242). The greenbelt around Seoul Metropolitan area is 1567 km^2 in 16 cities and 8 counties, including major cities like Seoul and Incheon (Benston and Youn 2004). And that of within the city of Seoul reaches 145 km^2. 64% of greenbelt areas are forests, 14% are farmlands, 4.4% are residential areas, and 16.3% are other uses such as roads, public facility sites, etc (Lee 2008).

2. Strict Regulation

After the greenbelt was designated, it was strictly maintained. In the greenbelt, land use conversion, installation of equipment, or new building construction was strictly prohibited. The only exception was the alterations and reconstruction of existing structures, of course, under government permission. For example, an existing housing unit may be expanded up to a total floor area of 33 pyong (100m^2, 1,111ft^2) with government permission, but it may not be subdivided into different units (Kim 1993, 67).

The main reason the greenbelt in the SMA has been managed so rigidly can be found in the political and administrative areas. The greenbelt was introduced by the central government and was specifically and strongly supported by the President. Throughout the

---

[8] The basic land use control system in Korea is the zoning. City Planning Act describes "specially designated zones": the Special Facility Control zone, the Development Restrict zone, the Anticipated Urban Development zone, and the Street Adjustment zone. Among them the Development Restrict zone is, so called, Greenbelt.

[9] City Planning Act, Article 21, Clause 1: "The Minister of Construction can designate Development Restriction Zone, when it is necessary to prevent sprawl of the cities, to create sound living environment for citizens by preserving natural environment of cities, and to limit the development of cities for the need of national security requested by the Minister of Defense."

1970s and 1980s, Korea kept their central planning system. So the central government has mainly been responsible for maintaining the boundary of the greenbelt. The role of local government was confined to enforce regulations and oversee any illegal building construction or conversion of land use in the greenbelt areas. As a result, although the central government is proud of keeping the original boundary of the greenbelt for three decades, residents and localities have complained about the unreasonable boundary and the restriction of inhabitants' property rights.

Evaluation of Greenbelt Policy

1. Development or Preservation

An urban greenbelt is a commonly used growth management tool for containing the physical expansion of built-up areas and preserving green space for environmental and recreational purposes. Generally a greenbelt is considered to lie outside the urban fringe. At the time of the establishment of a greenbelt, the urban boundary may lie well within the inner boundary of the greenbelt. However, as the city grows, the built-up areas stretched to the boundary of greenbelt. As a result, new policy decisions need to be made to determine continued desirability of the greenbelt. This is a critical question for the city of Seoul, one of the most rapidly growing cities in the world.

Many economists argue that the greenbelt is not an economically efficient land use (Choi 1994; Fujita and Lee 1997). In addition, Seoul experienced an extreme shortage of affordable land for housing, and rapidly increased land price. These problems have provoked hot debates on the greenbelt. Economists believe that the greenbelt policy has unnecessarily increased housing costs by restricting the expansion of housing construction into the greenbelt. However, most environmental planners have different opinions (Benston and Youn 2004; Kwon 2004; Lee 2008). They argue that the greenbelt has prevented urban sprawl and functioned as a source of clean air and amenity, even though the value of the greenbelt may not have been detected in monetary terms in economic analyses. And the general public strongly supports this idea.

The importance of greenbelt, which was drawn up to preserve nature and therefore maintain green zones under the ever-strengthening threat of air and other environmental pollution, in this era of industrialization and urbanization, cannot be overemphasized.

2. Negative Effects of Greenbelt

Greenbelt areas circled around Seoul were first set up in the 1970s, and have been well preserved. Although the greenbelt has many positive effects on the preservation of urban open spaces as well as the control of unregulated development in suburbs, however, there are also negative aspects on the inflexible applications and on the restriction of the right of resident's property that bring about many inconveniences. This section describes the negative effects of greenbelt in more detail.

First, the development of private properties in the greenbelt has been strongly restricted but proper compensation was not provided. Residents in the greenbelt have observed the escalating urban development and skyrocketing land prices. According to surveys (Ha and Cho 2009; Kim 1991), most residents expressed their dissatisfaction with the existing greenbelt regulations and were particularly critical of an apparent inequity resulting from what they regarded as an arbitrary practice of greenbelt boundary designation. More and more residents became disenchanted with the current laws which prohibited the alteration or extension of existing housing structures.

When the greenbelt was established, there was no compensation program for the landowners and the residents in the greenbelt area. However, the taking issue was not raised at all under the dictatorship during the 1970s. It was the time of social tension: "Yunsin" dictatorship, strained confrontation between North and South Korea, anxiety on national security with dispute on the withdrawal of U.S. military forces, and etc. However, in the 1980s, after the collapse of the military regime, and along with the democratization movement, the greenbelt became a political agenda. More than 80% of the greenbelt lands are privately owned property, and owners could not fully exercise their rights. However, neither central nor local governments were able to provide enough money to compensate their all losses.

Secondly, the greenbelt regulations are equally stringent to all of the lands without flexibility taking the site-specific conditions into account. There is a critical need to design different types and levels of regulations, considering the local information such as site characteristics, current land use, future demand of development, and environmental sensitivities of each tract of lands.

Third, when the greenbelt was established, there was limited survey data (research) available and less attention was given to the studies of the effects of the greenbelt on urban development and natural resources. So the local characteristic of each site was not well taken. Because of the centralized planning system, the government dominated the planning process and the stakeholders' participation was very limited. As a result, the boundary line was unrealistic. For example, the boundary line runs through the middle of 49 buildings and 38 built-up areas, all of which existed when the greenbelt was created (Lee 1994). In addition, there are virtually urbanized residential districts in the greenbelt where many residents have non-agricultural jobs outside greenbelt.

Fourth, the potential of recreational use and environmental amenity is under-utilized. After the designation of the greenbelt, it was strictly maintained and well preserved. However the functions and uses were not properly designed. Major parts of the greenbelt areas have been nearly abandoned without proper care and adequate utilization, because most areas are forested and landowners cannot exercise property rights under present regulations. As the greenbelt has a large amount of land resources and comprises 27% of Seoul's total areas, some parts of the greenbelt should be utilized for citizens' recreational areas and productive forests or farmlands.

Major open spaces of Seoul are the greenbelt, the Han-river and public parks. Most open spaces are located in suburban area. The inner city open spaces scattered around the downtown are relatively very small and few, and even more the distribution is not well organized in terms of urban open space system. Therefore, the greenbelt is vital to secure a less polluted environment, and also to provide various urban recreational activities.

APPLICABILITY OF TDR

TDR has been called a "win-win" solution because it allows both the public sector and the private sector to achieve their goals. TDR provides the compensation that makes it possible to adopt the land use restrictions needed to preserve important community resources. In Seoul's greenbelt case, the application of TDR may present a possibility of rearranging the boundary of greenbelt, protecting environmental resources, and compensating the loss of landowners and residents in greenbelt. TDR can also be helpful to ease the heavy financial burden on the deficit-ridden city government

Can TDR be successful in Seoul's greenbelt case? From the U.S.'s experience, prerequisites for a successful TDR programs are drawn. With these guidelines, this section analyzes the possibilities of TDR application to the greenbelt area of Seoul.

1. Greenbelt and TDR

TDR shares some features of the greenbelt program. It achieves the same clustered development patterns within fixed growth areas (receiving areas) and limits new development outside the growth area. However the most striking difference is that TDR shares the increased land values inside the greenbelt boundary with landowners excluded from it and who are prevented from future development.

2. TDR Possibility and Issues

Seoul's greenbelt case has suitable conditions for the successful TDR application. First, there exist clear goals and strong public acceptance. To be a successful TDR, there should be a clear goal and rationale for the program and the justification for the exercise of the police power (Keene 1997). And the need for resource protection should be supported by the community (Machemer and Kaplowitz 2002). Public disapproval can be a barrier to the effective implementation of TDR programs. Particularly at the local level, TDR programs need to obtain community support through consensus building and educational activities.

Seoul's greenbelt has the clear goal of protecting green spaces and ecosystems. In an urban environment, green space is indispensable for the well-being of residents because of its diverse ecological functions such as air pollution filtration and the conservation of biodiversity. It also provides amenity benefits related to scenic beauty and recreational opportunity. It is very important for the Seoul metro region's environmental quality. Therefore, Seoul greenbelt program has enjoyed a strong public support from politicians, nonprofit organizations, environmentalists, planners as well as the general public. According to a survey (Kim 1993), 90% of people answered that the greenbelt should be preserved, and 65% chose environmental protection for the main function of greenbelt while 10% chose controlling urban sprawl.

Second, a successful program requires sufficient demand for development. One important factor contributing to the failure of many TDR programs is that developers have little incentives to purchase development rights/credits. At the heart of TDR lies a presumption that there are either actual or potential pressures for growth. If there is no growth, land will be valued at its use value, and because there is no additional speculative value, its development value will approach zero. Consequently, it would appear that only in areas of development pressure would TDR be appropriate.

In the case of Seoul's greenbelt, there exists a strong development pressure. With the low land price[10] compared to the adjacent non-greenbelt lands, the shortage of housing units and the lack of developable lands, Seoul would assure the development pressure concentrated in receiving areas, which is essential for a successful TDR program.

While there are many positive conditions for the application of TDR as shown above, there also exist some concerns. The first possible issue is 'the unfamiliarity with TDR'. American experience suggests that public education and involvement is critical to achieving the ultimate goal of TDR programs. Because TDR is novel and sophisticated, the community must be prepared to engage in explanation and adjustment of the details of the program.

The second possible issue is 'intergovernmental relations'. TDR programs involving more than one municipality can be easily envisioned. The development from one community is transferred to another better suited to accommodating growth. Seoul Metropolitan Area's greenbelts are related to Seoul, KyungKi-do and Inchon. Therefore, intergovernmental cooperation and a comprehensive plan will be needed.

The third possible issue is 'equity' issue. Support and opposition to TDR programs will not be consistent from one region to another, and the interests of stakeholders, such as the municipality, developers, farmers, and residents, are different. Land owners in preserving areas and thus excluded from the receiving areas would probably protest the TDR program as unfair, even though they will be compensated by selling development rights. Because the long term land price increase in the receiving areas may be expected to far exceed that of sending areas after the transfer of development rights.

3. Suggestions for the Design of TDR Application

The first step in creating a TDR program is identifying the resource to be preserved. For the environmental protection in greenbelt area, it is critical for those people who will be affected by the TDR program to understand the goal and agree on the need to take action. If the community agrees that the protected areas are critical resources that contribute to the quality of life, the community is more likely to coalesce behind the TDR program and overcome the socio-political inertia that impedes the implementation of such an innovative program (Stinson 1998). To encourage citizen participation, coordinate interests of stakeholders, and achieve agreements, the objective, reliable and detailed information is needed (Kwon 1992; Kwon 2004). The resource targeted for the protection should be objectively surveyed through the environmental evaluation of greenbelt (Lee 2008). When it was first designated, due to lack of detailed information, the boundary was drawn absurdly. After completing the detailed analysis, the authorities need to draw a new layout, mapping out

---

[10] According to Choi's (1993) research, there is a structural shift between greenbelt and non-greenbelt values with the former being significantly lower than the latter regardless of land categories. It is estimated that the greenbelt land values are typically around 70% of the non-greenbelt land values, though the exact proportion varies in the range of 60% to 870% depending upon land categories. This implies that government restriction on land use in its most extreme form can lower the value of land by about 30% against market pressure for development.

regions under protection and free for public use. This new layout means the rearrangement of greenbelt areas, reclassification of land areas as well as a boundary redefinition based on each parcel's site-specific needs. When they design the new figure, they need to adopt clear and fair criteria, and include the consensus building of the community.

The next step is the design of the sending and receiving areas. While there are many arguments about the reclassification of greenbelt areas, I roughly categorize them into four areas, such as absolute preservation area, preservation for the future development area, residential development area, and recreational development area. The first category is the absolute preservation area. Because of strong development pressure, part of greenbelt can be encroached. So strict restrictions are still needed for the preservation areas of greenbelts such as drinking water resource areas, environmentally sensitive areas, biodiversity areas, and natural forest areas. These areas will be sending areas. The second is the preservation for the future development area. It is related with the concept of sustainability. These areas are preserved for the next generation, and designated as the sending areas. The third is the recreational development area. The city of Seoul does not contain enough recreational facilities. The greenbelt areas have large amounts of land resources for citizen's potential recreational activities such as a physical training field, green walk ways, bike ways, etc. These areas are receiving areas. The fourth one is the residential development areas. 7.3 km^2 (4.5% of Seoul's greenbelt) areas were occupied by residential housing units before the greenbelt designation. And another 27 km^2 (16%) are now occupied by public facilities. These areas can be developed within similar density level of nearby outside-greenbelt areas. These areas are the receiving areas.

TDR programs can be either voluntary or mandatory. Voluntary programs provide landowners the option of transferring development rights or developing their land. However, mandatory programs require developers to purchase rights in order to develop over the low base density. These programs also restrict landowners in the sending area from developing on their land. From a political standpoint, a voluntary program tends to be appealing to many landowners because they feel less constrained in their development options (Gottsegen 1992, 106). As a result, landowners are more likely to express support for a program that gives them another option for development but does not restrict their current choices. Thus, elected politicians and municipal governments may prefer the less political burden of voluntary programs over mandatory ones. However the American experiences illustrate the weakness of voluntary programs. They do not prevent new development from occurring in the area targeted for preservation, and often do not encourage development in the receiving areas. For example, the Collier county TDR program in Florida sought to protect sensitive everglades and big cypress preserve lands. However, since the TDR program is voluntary, it regulated but did not prohibit development in the sending area (Mittra 1996). This resulted in the environmentally sensitive region being developed. In contrast, the mandatory programs (Montgomery county, Maryland) limited the development in the sending area by strict regulations on building.

In conclusion, mandatory programs are likely to achieve preservation goals. These can be resulted from the overwhelming public support for the preservation of a particular resource. For the greenbelt of Seoul, there is a strong public support of preserving greenbelt. So the mandatory program seems to be likely to achieve its goal.

CONCLUSION

To manage urban growth and protect environmental quality, two public policies, TDR and urban greenbelt, are discussed in this chapter.

TDR is a complex, but useful and innovative growth management program for accommodating preservation and development simultaneously. TDR also serves as a tool for relieving some of the financial burden on property owners whose land has been legitimately downzoned. These TDR programs have been evolving in variety of land use situations, overcoming many initial questions posed on the earlier programs through the several decades' experiences of the United States.

The TDR program is particularly applicable to Seoul's greenbelt case. It has successful conditions for TDR. Seoul has clear goals of controlling urban growth, and protecting green spaces and ecosystems. It also has great support from general public, and sufficient demand for the development. The application of TDR appears to offer advantages to local governments that want to manage land use, re-adjust the greenbelt areas, protect ecological resources, and compensate land owners for restrictions on the development potential of their properties. However, it should be noted that each TDR program has different goals and operates in different environmental and political climates. Therefore, the planners who work with TDR programs have to deal with the implications of these differences. TDR designs require an understanding of development demands and patterns in order to appropriately locate sending and receiving areas as well as densities. Even though TDR is still a new concept in Korean planning practice, it is worth paying more attention to the application of TDR concept and the development of programs that can be acceptable to Seoul's planning situation.

Greenbelt may play an important role for the realization of eco-city functions. Thus the greenbelt policy should be approached by the mechanism of ecological conservation rather than the restriction of development. Regarding the rearrangement of the Seoul's greenbelt policy, planners need to consider environmental sensitivity, ecological approach, citizen participation, and equity to facilitate sustainable growth as well as to preserve the ecological nature of the greenbelts that enhance well-being of residents.

REFERENCES

Ahn, Gen-Hyeok and Yeong-Tae Ohn. 1997. A Critical Review on the Urban Growth Management Policies of Seoul since 1960. *The Journal of Korean Planners Association.* 32(3)

Ali, Amal. 2008. Greenbelts to contain urban growth in Ontario, Canada: Promises and Prospects. *Planning, Practice & Research.* 23(4): 533-548.

Amati, Marco and Makoto Yokohari. 2007. The Establishment of the London Greenbelt: Reaching Consensus over Purchasing Land. *Journal of Planning History.* 6(4): 311-337.

Barrese, James. 1983. Efficiency and equity considerations in the operation of transfer of development rights plans. *Land Economics*, 59(2): 235-296.

Bengston, David and Youn Yeo-Chang. 2004. Seoul's Greenbelt: An Experiment in Urban Containment. Proceedings of a symposium at the society for conservation biology 2004 annual meeting.

Chavooshian, Budd and Thomas Norman. 1973. Transfer of Development Rights: a new approach in land use management. *Urban Land.* 73: 11-16.

Choi, MackJoong. 1993. *Spatial and Temporal Variations in Land Values: A descriptive and behavioral analysis of the Seoul Metropolitan Area*, Ph.D. dissertation, Harvard University.

Chomitz, Kenneth. 2004. Transferable Development Rights and Forest Protection: An Exploratory Analysis. *International Regional Science Review.* 27(3): 348-373.

City of Seoul, *City Planning of Seoul for 2000s*, Seoul: City of Seoul, 1990. (in Korean).

Costonis, John. 1973. Development Rights Transfer: An Exploratory Essay. *Yale Law Journal.* 83(1): 75-128.

Fiedl, Barry and Jon M.Conrad, 1975. Economic Issues in programs of Transferable Development Rights. *Land Economics.* 51(4): 331-340.

Fujita, M. and C.M. Lee. 1997. Efficient configuration of a greenbelt: theoretical modeling of greenbelt amenity. *Environment and Planning A*, 29(11). 1999 – 2017.

Fung, Felix and Tenley Conway. 2007. Greenbelts as an Environmental Planning Tool: A Case Study of Southern Ontario, Canada. *Journal of Environmental Policy and Planning.* 9(2): 101-117.

Gans, Ellis. 1975. As a Method of Avoiding the Windfalls and Wipeouts Syndrome. in Jerome Rose, (eds.), *The Transfer of Development Rights: A New technique of Land Use Regulation.* N.J.: Center for Urban Policy Research.

Gottsegen, Amanda Jones. 1992. *Planning for Transfer of Development Rights: A Handbook for New Jersey Municipalities*, Morristown, N.J.: The New Jersey Conservation Foundation.

Ha, Seong-Kyu and Seong-Chan Cho. 2009. Suburban Development and Public Housing Provision on Greenbelt Zones in the Seoul Metropolitan Region. *Journal of Urban Administration.* 22(1): 183-207.

Henderson, Harold. 1998. Saved by Development: Preserving environmental areas, farmland and historic landmarks with Transfer of Development Rights. *Planning.* 64(3): 29-31.

Heui-Yeon Hwang. 1991 Greenbelt Regulation and Policy Alternatives. Korean Space and Environment Group, (ed.), *Space and Society*, Seoul: Space and Environment. (in Korean).

Johnston, Robert and Mary Madison. 1997. From Landmarks to Landscapes: A review of Current Practices in the Transfer of Development Rights. *Journal of the American Planning Association.* 63(3): 365-378.

Kahn, Sanders. 1984. Zoning and Transfer of Development Rights. *The Appraisal Journal.* 49(4): 556-563.

Keene, John. 1997. *Transferable Development Rights and Farmland Protection*, notes for the use of students.

Kelly, Eric. 1993. *Managing Community Growth*, London: Praeger.

Kenworthy, Jeffrey. 2006. The eco-city: ten key transport and planning dimensions for sustainable city development. *Environment and Urbanization.* 18(1): 67-85.

Kim, JooChul and Sang-Cheul Choe. 1997 *Seoul: the making of a metropolis*, NewYork: Wiley.

Kim, JooChul. 1991. Urban redevelopment of greenbelt-area villages: Study of Seoul, Korea. *Bulletin of Concerned Asian Scholars*. 23.

Kim, Jyung-Hwan. 1993. Housing Prices, Affordability, and Government Policy in Korea. *Journal of Real Estate Finance and Economics*. 6(1): 55-71.

Kim, Tae-Bok. 1993. A Study on the Management of Green Belt in Korea – Focused on the comparison with Great Britain. Seoul: GunHwa.

Koppel, Bruce and D.Young Kim. 1993. *Land Policy Problems in East Asia*, Seoul: Korea Research Insttute for Human Settlements.

Korea Times, April 7, 1998.

Korea Times, August 30, 1996.

Kwon, Kiewok. 1996. *Transfer of Development Rights: Evolution and Experiences in the United States and Exploratory Proposals for Seoul, Korea*, Unpublished Professional Project. University of Pennsylvania.

Kwon, WonYong. 1992. The Problems of Greenbelt and Policy Alternatives. presented at Greenbelt Policy Conference, Seoul, July 27, 1992. (in Korean).

Kwon Youngwoo. 2004. Regional Policy after the Deregulation of Greenbelts in Korea. *Geographical Studies* 38(3): 241-258. (in Korean)

Lee, C.M. and P. Linneman. 1998. Dynamics of the Greenbelt Amenity Effect on the Land Market: the case of Seoul's Greenbelt. *Real Estate Economics*. 26(1): 107 - 129.

Lee, Chang-Moo. 1994. *Greenbelt Impacts on Dynamics of Physical Urban Development and Land Market: A Welfare Analysis – The case of Seoul's Greenbelt*, unpublished Ph.D. Dissertation, University of Pennsylvania.

Lee, Chang Seok, Anna Lee and Younchan Cho. 2008. Restoration Planning for the Seoul Metropolitan Area, Korea. in Margaret Carreiro, Young-Chang Song and Jianquo Wu. (eds). *Ecology, Planning, and Management of Urban Forests: International Perspectives*. Springer.

Lee, GunYoung. 1995. *Cities and Nation: Planning Issues and Policies of Korea*, Seoul: Nanam Publishing House.

Lee, GyongJae. 2008. Long term improvement of development restriction zone of Seoul. *Urban Problems. 42-55.* (in Korean).

Lee, Jong-Il. 1987. Transition of the Concept of the Green Space and Property Rights. *Urban Problems*. 32. (in Korean).

Levinson, Arik. Why oppose TDRs?: Transferable development rights can increase overall development. *Regional Science and Urban Economics*. 27: 283-296.

Levy, John. 2008. *Contemporary Urban Planning,* 8th edition, Upper Saddle River, N.J.: Prentice Hall, Inc.

Lyu, Dong-Ju, 1987. Deed Restrictions on Greenbelt lands and Utilization Proposal. *Urban Problems*. 32 (in Korean).

Machemer, Patricia and Michael Kaplowitz. 2002. A Framework for evaluating Transferable Development Rights Programmes. *Journal of Environmental Planning and Management*. 45(6): 773-795.

Maryland Office of Planning. 1995. *Managing Maryland's Growth: Models and Guidelines,* Maryland: Maryland Office of Planning.

Mittra, Maanvi. 1996. The Transfer of Development Rights: A promising tool of the future. Land Use Law Center, Pace University School of Law. Available at http://www.law.pace.edu/landuse/tdrpap.htm.

Na, Hye-young, Byungseol Byun and Youngwoo Kwon. 2005. Case Studies on the Green Belt in the United Kingdom and the United States. *Geographical Studies* 39(1): 121-131. (in Korean)

Nelson A.C. and J.B.Duncan. 1995. *Growth Management Principles and Practices*, Washington, D.C.: APA Planners Press.

Nelson, Arthur. 1985. Demand, Segmentation, and Timing Effects of an Urban Containment Program on Urban Fringe Land Values. *Urban Studies*, 22(5): 439-443.

Nelson, Arthur. 1992. Improving Urban Growth Boundary Design and Management. *Real Estate Finance.* 8(4): 11-22.

Pedowitz, J. 1984. Transferable Development Rights. *Real Property, Probate and Trust Journal,* 19: 604-624.

Peters, James. 1990. Saving Farmland: How Well Have We Done?, *Planning,* 56(9): 12-17.

Pizor, Peter. 1978. A Review of Transfer of Development Rights. *The Appraisal Journal* 13(9): 386-396.

Pizor, Peter. 1986. Making TDR work: A Study of Program Implementation. *Journal of the American Planning Association.* 52(2): 203-211.

Poole III, Samuel E. 1984. TDRs in Practice: The New Jersey Pinelands. *Urban Land.* December.

Price, Dale. 1984. An Economic Model for the Valuation of Farmland TDRs. *The Appraisal Journal,* 49: 547–556.

Pruetz, Rick. 1996. *Putting Transfer of Development Rights to work in California,* Point Arena, CA: Solano Press.

Pruetz, Richard. 1997. *Saved by Development: Preserving Environmental Areas, Farmland and Historic Landmarks with Transfer of Development Rights, Burbank, Cal.: Arje Press.*

Raymond, George. 1981. Structuring the Implementation of Transferable Development Rights," *Urban Land*, 81: 19-25.

Rose, Jerome. 1975. *The Transfer of Development Rights: A New Technique of Land Use Regulation*, N.J.: Center for Urban Policy Research.

San Luis Obispo County, Department of Planning & Building. 1996. *Transfer of Development Credits.*

Sax, Joseph. 1971. Taking, Private Property and Public Rights. *The Yale Law Journal*, 81(2): 149-186.

Seoul Metropolitan Government. 1991. *Urban Planning of Seoul 1394-1994*, Seoul: Seoul City.

Stinson, Joseph and Michael Murphy. Transfer of Development Rights. 1998. http://www.law.pace.edu/landuse/tdr.htm

Tae-Bok Kim. 1993. *A Study on the Management of Green Belt in Korea – Focused on the Comparison with Great Britain*, Seoul: GunHwa. (in Korean).

Taintor, Rick. 2001. *Transfer of Development Rights Report.* South County Watersheds Technical Planning Assistance Project.

Tang, Bo-sin, Siu-wai Wong and Anton King-wah Lee. 2007. Greenbelt in a compact city: A zone for conservation or transition?. *Landscape and Urban Planning.* 79: 358-373.

Thomas, David. 1970. *London's Green Belt*, London: Faber and Faber.

Tustian, Richard. 1983. Preserving Farming through Transferable Development Rights: A case study of Montgomery County, Maryland. *American Land Forum Magazine,* summer. 63-76.

Woodbury, Steven. 1975. Transfer of Development Rights: A New Tool for Planners. *Journal of the American Institute of Planners*. 41(1): 3-14.

Yeom, Heing-Min. 1993. A Special Issue: On the Relaxation of the Greenbelt Regulation. *Urban Information Service*. 12. (in Korean).

Ziegler, Edward H. 1995. The Transfer of Development Rights. *Zoning and Planning Law Report,* 18(8).

List of Cases

Dutour v. Montgomery County, Law Nos. 56964, 56968, 56969, 56970, and 56983, Circuit Court for Montgomery County, Jan. 20, 1983.

Fred F French Investing Co. V. City of New York, 39 N.Y. 2d 587; 385 N.Y.S. 2d 5 1976.

Lucas v. South Carolina Coastal Council, 112 S. Ct. 2886, 120 L. Ed. 2d 798, 1992.

Penn Central Transportation Co. V. City of New York, 438 U.S. 104, 1978.

Village of Euclid v. Ambler Realty Company, 272 U.S. 365, 1926.

In: Eco-City and Green Community
Editor: Zhenghong Tang

ISBN: 978-1-60876-811-0
© 2010 Nova Science Publishers, Inc.

Chapter 6

DESIGN THINKING: A POTENTIAL PLATFORM FOR THE 'REFLECTIVE PRACTITIONER AND PRACTICAL SCHOLAR' TO SPEAK?

Erin Bolton

Community and Regional Planning Program
University of Nebraska, Lincoln, NE, USA

"The world we have made as a result of the level of thinking we have done thus far creates problems we cannot solve at the same level of thinking at which we created them."
Albert Einstein

"If you look at the science about what is happening on earth and aren't pessimistic, you don't understand data. But if you meet the people who are working to restore this earth and the lives of the poor, and you aren't optimistic, you haven't got a pulse. What I see everywhere in the world are ordinary people willing to confront despair, power, and incalculable odds in order to restore some semblance of grace, justice, and beauty to this world."
Pawl Hawken
2009 Commencement Address, University of Portland

INTRODUCTION: CONFUSION AND BICKERING IN PLANNING

As a planning student, I am the proverbial child of two divorced, bickering parents – the planning academic and the planning practitioner. As a result of this separation, there appears to be conflict and confusion in planning (Fainstein and Cambell, 2003).

Fainstein et al. (2003) state that the practical scholar and the reflective practitioner ought to at least be able to speak to one another. Well, just as David C. Perry (1995) suggested a spatial approach as "a mode of thought" for planners to "make space," I'm proposing design thinking not only as a "mode of thought," but one that provides common ground for both the practical scholar and the reflective practitioner to discuss, evaluate, and envision better, or more just and sustainable, cities (Fainstein 2000, 2005; Campbell, H. 2006). To do so, I note the desires of planning theorists in *Readings in Planning Theory* (Campbell et al., 2003) and show how design thinking satisfies, creates, or makes room for them. I feel this is important to the because perhaps if practitioners and theorists could talk and interact more efficiently, they may have a greater impact on important issues of sustainability (i.e. addressing inherent conflicts underlying sustainability outlined in Campbell (1996).)

DESIGN THINKING

Design thinking is not a new, revolutionary idea. Tim Brown, CEO and president of design and innovation consulting agency IDEO, has popularized design thinking for the business world (see Brown, 2008). Stanford's new, multidisciplinary graduate design school, called the d.school, has set out to "create the best design school. Period" because they believe "great innovators and leader need to be great design thinkers" (see Stanford d.school, 2009). Bryan Lawson, psychologist and architect, has written a seminal book conceptualizing design thinking and problem solving entitled, "How Designers Think: The design process demystified" (2006). I rely heavily on his text for a number of reasons. Most importantly is because Lawson is both a psychologist and an architect. I find his cognitive approach empirically and conceptually fascinating. He forgoes the frequently bombastic nature of architectural discourse on design and instead addresses design thinking in an inclusive manner. He has taught students of traditional design mediums (i.e. architecture and urban design) as well as those not currently receiving a design education (i.e. town planning and computer sciences). So, what is design thinking? And, finally, how does it have room to bolster the outcries of important planning theorists and satisfy the impatience of practitioners?

Relationships between Problem and Solution

Design thinking is not easy to clarify and thus is not intended to be descriptive (Lawson, 2006), as, say, the rational method of planning. Lawson describes the design thinking as a three-dimensional process of "negotiation between problem and solution through the three activities of analysis, synthesis, and evaluation." He is careful to note that any diagram or prescription presents a danger of oversimplification of "what is a highly complex mental process" (Lawson, 2006, p. 49) and that designers for the most part learn how to define problems by trying to solve them (Lawson, 2006, p. 55). Brown (2008) likens design thinking to a series of incremental steps, or spaces.

According to Lawson (2006, p. 43), the scientist has a problem-focused thought process, while the designer has a solution-focused process. Christer Bengs, in "Planning Theory for the naïve?," echoes this concept for planning. Bengs contends that both planning theory and practice aim to solve problems as a response to alleged procedural or subject matter problems. He distinguishes this process from the rest of academia, which, he claims, attempts to *define* problems and from architecture and engineering, which he note as being solution focused.

Planning as noted in literature is normative and ethical by nature (Campbell, H. 2006). Bengs continues, "Moral dilemmas may of course concern problem formulation as well as problem solving."

So, then, if what Lawson and Bengs say is correct, design thinking would be educationally beneficial to planning students because it focuses on solving problems as well as formulating them.

Modernism and Postmodernism in Relationship to Design Thinking

In Beauregard's (1989), 'Between Modernity and Postmodernity: The Ambiguous Position of U.S. Planning,' he "offers suggestions as to how planning practitioners and theorists might respond critically to a postmodern capital restructuring and cultural transformation, heeding it's call to be flexible and open but not abandoning the modernist quest for a democratic and reformist planning and a commitment to the city." Essentially, he is calling for planning to take the best from both camps. From the modernist camp, he calls for a commitment to the city, to reform, and to democracy. I contend that design thinking propagates that call using Bryan Lawson's work. Design thinking is a priori considerate of the physical city form, but what else does it offer in regards to "Between Modernity and Postmodernity?"

Design Thinking Committed to Reform?

Lawson (2006, p. 181) notes that not only is it very common for guiding principles to be carried throughout a designer's life, but that "Designers work in the context of a need for action. Design is not an end in itself. The whole point of the design process is that it will result in some action to change the environment in some way, whether by the formulation of policies or the construction of buildings. Decisions cannot be avoided or even delayed without the likelihood of unfortunate consequences" (Lawson, 2006, p. 125).

Design Thinking Is not Technocratic

Design thinking encourages planners to avoid technocratic methodologies. Like the technocratic planner, Lawson (2006, p. 71) states that "The unthinking designer could easily use such an apparently high quality and convincing data to design an office based on such factors as minimising 'person door movements'. [...] such figures are quite useless unless the designer also know just how important it is to save 7 seconds of time."

Design Thinking Committed to Democracy?

Design thinking strongly adheres and respects the needs of client and user (Lawson 2006). Lawson (2006, p. 84) writes that, "In design, the problem usually originates not in the designer's mind but with a client; someone in need who is unable to solve the problem, or perhaps, even fully understand it without help." With that being said, he notes that the relationship between designer and client constitutes much of the design process. I see this notion to be appropriate to planning in the follow analogy: Designer: Client; Planner: Citizen.

Design thinking is flexible and open because, "...there is no correct 'method' of designing, nor one route through the process" (Lawson, 2006, p. 200). "This [legislation making design more difficult] is not because it imposes standards of performance which may be quite desirable, but because of the inflexibility and lack of value which it introduces into a value-laden multi-dimensional process which is design" (Lawson, 2006, p. 75).

Design Thinking on Neutrality and Inaction

Beauregard (1989, p. 121) concludes with this notion: "Instead of hiding behind the cloak of expertise or remaining distant from controversy, planners must participate." Lawson (2006, p. 115) identifies the need for action within design thinking: "Procrastination as a strategy is deeply flawed. In many real-life design situations it is actually not possible to take no action. The very process of avoiding or delaying a decision has an effect!"

In "Toward a Non-Euclidian Mode of Planning," Friedmann (1993) writes and argues a list of qualities planning should have. Planning, according to Friedmann, should be: normative, innovative, political, transactive, and based on social learning.

Lawson Writes that Design Thinking Is Normative in Nature

"While scientists may help us understand the present and predict the future designers may be seen to prescribe and to create the future, and thus their process deserves not just ethical but also moral scrutiny." (Lawson, 2006, p. 125).

Design Thinking Creates Innovation

"...designers rely on information to decide how things might be, but also they use information to tell them how well things might work" (Lawson, 2006, p. 119). This fact about design thinking tells me that it is inherently innovative. According to Tim Brown (2008), design thinking, via a focus on human-centered design, inspires innovation. He writes, "...innovation is powered by a thorough understanding, through direct observation, or what people want and need in their lives..." (Brown, 2008, p. 1).

Design thinking can be considered political because it involves collaboration of different groups of people (i.e. designer, user, and client) and can examine integrated systems as asked of planning professionals: "Thus it is the case that good design is usually an integrated response to a whole series of issues. If there is one single characteristic which could be used to identify good designers it is the ability to integrate and combine" (Lawson, 2006, p. 62).

Design draws from expertise and experience perhaps because, "...reasoning and imaging were probably the most important to designers. Reasoning is considered purposive and directed towards a particular conclusion. This category is usually held to include logic, problem-solving and concept formation. When 'imagining, on the other hand, the individual is said to draw from his or her own experience, combining material in a relatively unstructured and perhaps aimless way" (Lawson, 2006, p. 137).

Design Thinking Is Based on Social Learning

Design thinking is communicative and experiential because, as aforementioned, it largely learns about problems by trying to solve them. In other words, a student learns design thinking through experience.

Additional Support

What about the 'piecemeal' approach and reality Lindblom (1959) speak of? Design thinking recognizes that efforts are imperfect (Lawson, 2006, p. 119) and actually warns against procrastination, which according Lawson, results when a designer falsely believes incredibly complex concepts will clear up if (s)he waits awhile.

Design Thinking Recognizes Traps (Lawson, 2006).

Jane Jacobs (1961) unapologetically attacks planners or architects who press images, such as grass on people. Also, she criticizes Harvard and MIT studio classes where students "who now again pursue, under the guidance of their teachers, the paper eercise of converting it into superblocks and ark promenades…" Adhering to a image, or an icon, without reason and context can be attributed to traps in design thinking (Lawson, 2006). However, because design thinking acknowledges those traps, it has the capability to self-check and address lapses in judgment, which is crucial in planning.

POTENTIAL EDUCATIONAL IMPLICATIONS

Like Stanford's design school, I would imagine design thinking for planning to be multi-disciplinary in much the same sense as practice. Planning would probably head back into studio educational formats. However, this studio cannot be a lab where students design in a vacuum. The studio would need to be conducted on a street-level, with students engaging within communities and also with the city itself. The embodiment of design thinking in planning would probably resemble Auburn University's Rural Studio in some way. Rural Studio, founded in 1992 by Samuel Mockbee, is the infamous design-build program that joins architecture students and community residents in Hale County, AL. Their mission is "to allow students to put their educational values to work as citizens of a community." They seek "solutions to the needs of the community within the community's own context, not from outside it" (see Rural Studio 2009). They build Boys & Girls Clubs, fire stations, 20K Houses, parks, and baseball diamonds – actually build them, not simply discuss plans to build them. *Metropolis* magazine (2009) hailed, "The legendary design-build program figures our affordable housing." Everything they are doing, aside from designing the buildings themselves, resembles planning, its processes, and its goals for social justice.

In addition, I could envision students focusing on concrete problem sets. For example, during Fall 2008, Illinois Institute of Technology had a focused class called, "Prototyping Chicago Bus Rapid Transit," where students used behavioral prototyping methods to improve seating layout on Chicago regional transit buses and presented their plans to the CTA (see ITT, 2008). Perhaps there would also be a "policy studio" where students learned how to debate and propose policy solutions regarding, say, case studies in the sense Paul Davidhoff (1965) advocated. This would distinguish a planning studio, say from a classic architectural studio, and would be incredibly important.

COMMENTARY: A STUDENT'S PERSPECTIVE ON PRAGMATISM AND PLANNING

"The idea that university people who train planners should be engaged in the design of a new society – at the same time as they absolutely reject the idea of any kind of blueprint design for any other dimension of life – is slightly absurd." Christer Bengs.

The European Journal of Spatial Development

In "Rethinking Planning Theory for a Master's Level Curriculum," Nancy Frank (2002) discusses the validity, objectives, failures and pedagogy of teaching planning theory to students. She compares syllabi and teaching methodologies and expresses the difficulties she has personally experienced teaching planning theory courses. Noting her trials, she writes, "Not surprisingly, students left the experience [course in planning theory] with no clear sense of what they should have learned or why they should have learned it." In all honesty, that statement embodies what I felt even before my planning theory course ended.

Further, Frank contends that theory courses should include serious critiques of postmodernism: "In other words, our own postmodernist leanings – our intense effort to avoid imposing our values on students – may get in the way of mentoring our students to adapt a practice that is both ethical and effective, without being authoritarian." As a student, I concur. I would like planning professors to teach me how to think. I am not saying they ought to tell me what to think, but they should not shy away teaching students the critical thinking skills that may guide them in the face of tough planning questions and decisions.

I had a statics engineering professor who unblushingly stood in front of the class and whipped through the mathematics necessary to complete assigned problems. Audible frustration encircled the room. Finally, once the orchestra of groans became too great to ignore, she turned and asked defiantly, "What is it you all are missing here?" and drew a new, although conceptually related, free body diagram on the board. She then paused, peering around the room and waited.

Meanwhile, students rushed to scribble the problem on their graph paper, fumbled to reset their calculators or, for those over-achievers, frantically began calculations. "Stop!" she commanded, "Put everything down and look at the problem. Can't you see what the answer is? Just look at it!"

This question evoked levels of frustration nearing anger from the class. How were we to know the outcome of a problem not yet calculated? Of course she did not expect for us to know the exact outcomes of calculations we hadn't completed, she merely asked that we use common sense and grasp what was happening generally. If the forces on one side of the beam exceeded those on the other, well, then beam would no longer be balanced or static. That's common sense. This is a gross over-generalization of engineering concepts, but I believe there is something useful in taking a step back, observing, and developing a general understanding of where there is imbalance.

As noted, I am a second year planning student with a background in liberal arts and design. I am finished with core courses on subjects such as planning theory, economics, GIS analysis, qualitative methods and quantitative analysis. I chose to study planning because throughout my time traveling and studying liberal arts I developed deep convictions regarding social justice and wanted to become an individual apt and knowledgeable at creating better - or more lively, just, and diversely beautiful - cities. What that meant specifically, I did not pretend to know. However, after a year of rigorous study, I am left more lost and confused about planning than ever before. Could this be a reflection of the identity crisis drenching the planning profession, not to mention the "feud" between planning academics and planning practitioners?

In the spirit of stopping and simply trying to see what is happening, perhaps I ought to begin with pragmatism. I do this because I see pragmatism as the most practical derivative of post-modernism.

Problems with Pragmatism

"In summary, then, by focusing entirely on the concrete and pragmatic, the urban design and planning professions have disassociated themselves from universal questions. Furthermore, this is happening at the exact moment when these questions are resurfacing."

Tali Hatuka and Alexander D'Hooge (2009)

When I read sociological and political theories, I was taught to begin with the following baseline questions: "What is this theorist saying about the role of the individual versus collective society? What then, about governance?" or "What is the root cause, according to this person, of behavior here? Social conditioning? Biology and evolution?" Subsequently, when I read overarching philosophical theories I start by asking the questions, "What is this philosophy saying regarding the nature of truth? Is it logically consistent? What are the implications?"

For example, a pragmatist states, "All truth is relative. There are no absolute givens!" (Campbell, H. & Marshall, 1999) And, in the same breath he announces, "Pragmatism is the correct way, not modernism!"

Here's a more tangible example. Chris and I are a couple, and we enjoy cooking meals together. Chris strictly adheres to the recipes - every teaspoon of salt is exact; the timer is set to cook pasta precisely. I, however, am a bit more haphazard. I may study the recipe beforehand, but I rarely set timers and measure very little. In fact, my measurement as to whether something is done in the oven is gauged by sight and aromatics rather than time. On a particular day, I spontaneously decide making a béchamel sauce is necessary for our meal. I know there is flour, milk, and butter involved but that is about it. This of course, in my mind, is okay, because recipes are not "my thing," anyhow. Chris, heeding my vocalized desire for béchamel sauce, quickly walks toward the computer to find a recipe.

"Chris, why do we need that?" I snap. "You always need your recipes, timers, and measurements," Grumbling, I say, "We shouldn't use recipes anyhow. Your consistence need for structure aggravates me."

He replies remorsefully, "Um, well, I just thought we could use parts of a recipe to…"

And I burst into laughter.

Why laughter? Because I realized, what I was doing was chastising his way, or structure, and at the same time arguing that he ought to do things my way, or my "non-way way." In other words, while his approach may be significantly more structured than mine, non-structure is still a structure.

What is important about this? As stated, the pragmatist begins with the notion that there are no absolute givens. Well, except the given that (s)he is correct and someone else, in this case the modernist, is wrong. So, then how can one argue on the basis that modernism is wrong without appealing to another way as being correct? Furthermore, how can the pragmatist state a foundational truth that a thinker ought always approach a set of problems pragmatically, but not recognize foundational truth? Frankly, (s)he cannot. Logic doesn't allow it just as it did not allow me to say Chris should not appeal to structure when I was in fact demanding he adhere to my non-structure structure. "The given is that there are no givens," or perhaps more accurately, "the truth is that there is no truth," while rhetorically playful, is self-defeating. It fails the laws of logic (specifically the law of non-contradiction). Truth, by definition, is exclusive.

Moreover, why debate if all truth is relative? Doesn't the fact that planning theorists debate in the first place prove they appeal to some standard of correctness they expect the other individual to understand (see C.S. Lewis)? Perhaps in order to progress, planning theorists ought to make a distinction between pluralism in the cultural sense (i.e. diversity), and pluralism as it pertains to relative values, or truths (see Lennox). What does this mean?

The difference between the need for diversity and cultural pluralism in planning versus relative, subjective values in planning is quite easily distinguishable in Paul Davidhoff's famous article "Advocacy and Pluralism in Planning." Although Davidhoff's arguments have been understood as relative/subjective in its focus of concern (Campbell, H. et al., 1999), I do not feel that is the point. I would argue that Davidhoff is advocating the importance of cultural pluralism (i.e. the need to give subjective interpretations of foundational truths a voice) but never states that all truth is subjective or that, because all truth is subjective, there is no such concept of truth, or in this case justice.

He writes:

"The idealized political process in a democracy serves the *search for truth* in much the same manner as due process in law. Fair notice and hearings, production of supporting evidence, cross-examination, and reasoned decision are all means at relative truth: *a just decision*. (…) The legal advocate must *plead* for his own and his client's *sense of legal propriety of justice*. The planner as advocate would *plead* for his own and his client's view of *the good society*" (emphasis added).

What are these advocates pleading in relationship to? Davidhoff appears to notice that evidences, cross-examination, fair notice, etc. are means at relative, or subjective, truth, but they are not truth itself. He is contending for due process in cultural pluralism and diversity.

Diverse viewpoints are to be given a chance to be heard, but he does not say that they should be given equal credence. In fact, Davidhoff acknowledges the search for truth and justice inherent in the process. He argues that each opposing viewpoint should plead its "sense of legal propriety of justice" or "view of the good society."

I am not a lawyer, but I image this: if a lawyer is presenting and advocating his/her case, (s)he is appealing to some type of justice standard; attempting to prove that his/her case reflects that justice standard more accurately than his/her opponent's. The concept of justice then, must somehow stand on its own (Fainstein 2005) for their to be a reason to debate and oppose diverse interpretations of it, let alone make decisions regarding each case's validity and likeness to it. In other words, Davidhoff acknowledges and advocates opposing viewpoints and subjective interpretations of justice (i.e. those that perhaps derive from cultural pluralism), but he is does not negate the foundational sense of justice itself nor a due process which presses toward it. In fact, he states, "The prospect for future planning is that of a practice openly inviting political and social values to be *examined and debated.*"

TRUTH BY ITS NATURE

"Without regard for objective truth, one should not expect justice."

John Lennox

Perhaps I need to articulate more accurately. I lifted the following from my apologetics research. It is the most concise, clear way of beginning to understand the nature of truth and, furthermore, justice I have encountered.

Truth is, by nature:

Non-contradictory – it does not violate the basic laws of logic.
Absolute – it does not depend upon any time, place or conditions.
Discovered – it exists independently of our minds; we do not create it.
Descriptive – it is the agreement of the mind with reality.
Inescapable – to deny its existence is to affirm it.
Unchanging – it is the firm standard by which truth claims are measured.

(McAllister, DATE).

I addressed the non-contradictory manner of truth earlier when speaking about pragmatism beginning with the statement: "The given is that there is no given" or "The truth is that there is no truth." Also, the inescapable property of truth is wrapped into the aforementioned contradictory truth statement upon which pragmatism, and post-modernism, is built. The statement must affirm truth in order to attempt or negate it, which of course is contradictory, and leaves truth affirmed. In Davidhoff, we saw an application of the unchanging quality of truth in terms of justice—that justice is a standard to which competing viewpoints are measured. What about the absolute and discoverable nature of truth, and therefore justice? These two qualities are of upmost importance and have, in my opinion, great implications for planning practice because justice is central to planning (Fainstein 2000, 2005; Campbell, H. et al. 1999; Campbell, H. 2006).

If justice is absolute in that it does not *depend* on any time, place or conditions, yet it exists independently of ourselves, shouldn't we need to look toward context for clues about it? In other words, it appears that stating truth is absolute, or rather acknowledging foundational truths, is not in fact stating context, history, and culture do not matter. It is simply stating that context is not both the means and the end. If anything, context matters greatly, as it has the ability to point us toward a justice "discovered," or a just decision as Davidhoff put it.

Approaching justice in planning this way allows a person to be, in a sense, firmly planted "between modernity and postmodernity" because it recognizes justice in its commitment to reform and democracy and respects flexibility and openness necessary for cultural pluralism to thrive (Beaugard, 1989). This humbles a person; it allows opposing interpretations and provides criteria for their evaluation. It requires that, while a person has some notion of truth and justice, they must go outside themselves to affirm it, or discover it; context matters, experience matters, democracy matters, philosophy matters, and the sciences matter. This process drives a person; it will undoubtedly require individuals to question, realize, and expose their own assumptions and compel action and decision, rather than neutrality and ineffectiveness attributed to pragmatism and other practices based on postermodernism (Campbell, H. 2006). Justice separated from objective truth, as it is in pragmatism, quickly becomes synonymous to personal convenience and doing what is "right" for the moment. What would that mean for sustainability?

PRAGMATISM ATTEMPTS TO HANDLE VALUE-BASED CLAIMS

Understanding truth, and justice, with the presupposition that it is discoverable and absolute means also that a practical person will ask the practical, common sense questions typically attributed to pragmatism. Christens (2009, p. 32) suggests planners utilize pragmatic inquiry to expose value-laden claims within New Urbanism and Suburban Decentralization. In order to juxtapose the two schools of thought, he suggests asking the following questions: "What are the beliefs that inform this approach? If we are to adopt these beliefs in specific instances, what are the values for which we are working? And, why should we – in this context – believe these values to be worthy of our actions (and beliefs)?" Frankly, these questions are practical questions and should be asked of any value-based system. They are not exclusive to pragmatism, nor should pragmatism be held on a pedestal simply because it poses the right questions.

Further, Christens (2009, p. 33) acknowledges pragmatic inquiry "cannot be expected to result in a conclusion that a given theory or practice is more true or even more useful in every set of circumstances." He continues, "Since many people experience elements of both urban design patterns [Suburban Decentralization and New Urbanism] as satisfying, a pragmatic view holds that no meta-critique can be leveled against these experiences."

So, while pragmatism permits me to ask practical questions and brings value-laden claims into view, it gives me nothing to critique and measure my ultimate decision between New Urbanism and Suburban Decentralization against. While Davidhoff argues opposing plans ultimately plead for their "view of the good city," pragmatism is left neutrally ineffective because it does not recognize any standard, such as the just or good city. After all, it was Davidhoff (1965) who firmly stated, "The planner should do more than *explicate* the values underlying his prescriptions for courses of action; he should affirm them; he should advocate for what he deems proper" (emphasis added).

Maybe students need to be asking more probing questions. If the scientific data, context, and need for sustainable cities are present (although they were probably always present), why is it so difficult for people to join the sustainable movement? Why must planners even need to advocate for diversity, sustainability, and justice? If context, empirical evidence, and sufficient need were all planners required to get a movement like sustainability truly off the ground it would have been done already. What holds people back?

I appeal to Davidhoff, yet again:

"*A city is its people*; their practices; and their political, social, cultural, and economic institutions as well as other things. [...] The contemporary thoughts of planners about the nature of individuals in society are often mundane, unexciting, or gimmicky. [...] Planners seldom go deeper than acknowledging the goodness of green space and the soundness of proximity of linked activities. We cope with the problems of the alienated citizen with a recommendation for reducing the time of the journey to work."
(emphasis mine)

Do planners think about people? In what way?

HUMAN NATURE

"Because psychology is about people, it cannot shirk the responsibility of dealing with fundamental questions about human nature. In general, its audience already holds certain views on these questions. Every age has its own conceptions – men are free or determined, rational or irrational; they can discover the truth or they are doomed to illusion. In the long run, psychology must treat these issues or be found wanting."

Ulric Neisser *Cognition and Reality*, 1976

If the purpose of planning is to build a plan, based on a city or a society, it too is about people and for all illustrative purposes, I could replace the word *psychology* in Neisser's above quote with *planning*.

There are planners who study person-environment-behavior relationships. Jack Nasar (see Nasar, Jack L. & Wolfgang F.E. Preiser, 1999) is the first to enter my mind. Specifically related to transportation planning, I think of Tommy Gärling (see Gärling, Tommy & Kay W. Axhausen, 2003) who has written articles pertaining to travel and habitual behaviors impeding mode shifts from private car use to sustainable transit. However, they are certainly not the norm.

Steven Pinker, author of *The Blank Slate: The Modern Denial of Human Nature*, makes a case for the importance of acknowledging human nature and, in one instance, points specifically toward planning. He addresses Le Corbursier's, "Radiant City" utopian vision on page 170:

"In Le Corbursier's Utopia, planner's would begin with a 'clean tablecloth'... and mastermind all buildings and public spaces to service 'human needs.' [...] It did not occur to

Le Corbursier that intimate gatherings with family and friends might be a human need, so he proposed large communal dining halls to replace kitchens. Also missing from his list of needs was the desire to socialize in small groups in public spaces, so he planned his cities around free-ways, large buildings, and vast open plazas, with no squares or crossroads in which people would feel comfortable hanging out to schmooze."

Pinker concludes that Le Corburier's realization of the "Radiant City," as reflected in Chandigarh and later Brasília, was an absolute failure. He writes, "Today both cities are notorious as uninviting wastelands detested by the civil servants who live in them."

Of course, looking toward human nature for guidance is not *that* simple. Professors seem quick to say that, generally, socialism "fails" because human beings, by nature, need incentive. What appears to be forgotten is that capitalism "succeeds" because it is so closely in tune with human nature (i.e. economic principles such as unlimited wants and desires). That is also true. But is capitalism more desirable simply because it appears more "successful" at speaking to human nature? A priori, I would think not, but I do not begin know. This is where the student turns to their planning professors for guidance (recall Frank, 2002).

PLEAD FOR EDUCATION

As Lucy (1988) made a case against APA ethical standards, perhaps the student ought to examine APA's "Skills of Successful Planners." They are listed on-line as follows:
Skills for Successful Planners

- Knowledge of urban spatial structure or physical design and the way in which cities work.
- Knowledge of plan-making and project evaluation.
- Mastery of techniques for involving a wide range of people in making decisions.
- Understanding of local, state, and federal government programs and processes.
- Understanding of the social and environmental impact of planning decisions on communities.
- Ability to work with the public and articulate planning issues to a wide variety of audiences.
- Ability to function as a mediator or facilitator when community interests conflict.
- Understanding of the legal foundation for land use regulation.
- Understanding of the interaction among the economy, transportation, health, and human services, and land-use regulation.
- *Ability to solve problems using a balance of technical competence, creativity, and hardheaded pragmatism.*
- Ability to envision alternatives to the physical and social environments in which we live.
- Mastery of geographic information systems and office software.

In some ways planning education fulfills this list (i.e. Mastery of geographic information systems). In most ways, in does not. This is of course not entirely unfortunate, but deserves consideration. APA advising students that "hardheaded pragmatism" (whatever that means) is not only a skill, but also a must-have skill for a "successful planner" is, honestly, inadequate for me. Also inadequate is the "ability to function as mediator or facilitator when community interests conflict."

I mentioned before that as a planning student I am a proverbial child of divorced, bickering parents. However, amidst this divorce, I am not necessarily asking for parental reconciliation, rather an education. What I have read indicates that, because planning is normative and political by nature, it does not have that luxury of sitting silent on matters of justice nor can it afford to either simplify or ignore the grand diversity and complexity of the human condition. Thus, as a planning student, my education must reflect this and not that of the confusion, detachment and identity crisis rampant in planning. In order to plan, I ought to explore the objective nature of justice in a beautifully pluralistic society. To do so, I need the thinking skills to identify, evaluate, and criticize the presuppositions and implications of normative arguments. Also, I must be given the applied experience in order to treat matters of social injustice with empathy, articulated argumentation, and technical skill, all while understanding the structure of political, historical, economical, and cultural contexts. It is a tall order, and perhaps too much, for a two-year graduate degree. At the very least, however, students should have the tools to move down the path described above and have the reflective autonomy necessary to confront difficult professional decisions.

Again I turn to Davidhoff: "As members of a profession charged with making urban life more beautiful, exciting, creative, and just, we have had little to say. Our task is to train future generation of planners to go well beyond us in its ability to prescribe the future urban life."

CONCLUSIONS AND CRITICISMS

One criticism could be that design thinking, although it may provide a new framework for thought, still fails to provide guidance in the sense that it will somehow let us know what to do in a specific context, place, and situation. That is correct, it doesn't. However, it does provide the environment and education conducive to thoughtful critique and examination of concepts such as justice, sustainability, ect. Design thinking is dynamic, yet it recognizes constraints; It is a place where the practical scholar and the practitioner can at least speak to one another and perhaps imagine a more sustainable city.

REFERENCES

American Planning Association. "What Skills Do Planners Need?" http://www.planning.org/onthejob/skill.htm. Accessed August 11, 2009.

Beauregard, Robert A. (1989). Between Modernity and Postmodernity: The ambiguous Position of U.S. Planning. In Scott Campbell & Susan Fainstein (Eds.). *Readings in Planning Theory* (2nd ed., 2003: 108-124). Malden, MA: Blackwell Publishing, Ltd.

Bengs, Christer. Planning Theory for the naïve?. *European Journal of spatial Development.* http://www. Nordregio.se/EJSD/ Accessed August 11, 2009.

Brown, Tim (2008). Design Thinking. *Harvard Business Review,* June 2008: 1-10.

Campbell, Heather (2006). Just Planning: The Art of Situated Ethical Judgement. *Journal of Planning Education and Research* 26:92-106.

—— & Robert Marshall. (1999). Ethical Frameworks and Planning Theory. *International Journal of Urban and Regional Research.* 23 (3): 464-478.

Campbell, Scott. (1996). Green Cities, Growing Cities, Just Cities? Urban Planning and the Contradictions of Sustainable Development. In Scott Campbell & Susan Fainstein (Eds.). *Readings in Planning Theory* (2nd ed., 2003: 435-458). Malden, MA: Blackwell Publishing, Ltd.

—— & Susan Fainstein (Eds.). (2003). Introduction: The Structure and Debates of Planning Theory. *Readings in Planning Theory* (2nd ed.). Malden, MA: Blackwell Publishing, Ltd.

Christens, Brain (2009). Suburban Decentralization and the New Urbanism: A Pragmatic Inquiry into Value-based Claims. *Journal of Architectural and Planning Research*, 26:1, 30-43.

Davidhoff, Paul. (1965). Advocacy and Pluralism in Planning. In Scott Campbell & Susan Fainstein (Eds.). *Readings in Planning Theory* (2nd ed., 2003: 210-223). Malden, MA: Blackwell Publishing, Ltd.

Fainstein, Susan. (2005). Planning Theory and the City. *Journal of Planning Education and Research* 25:121-120.

—— . (2000). New Directions in Planning Theory. In Scott Campbell & Susan Fainstein (Eds.). *Readings in Planning Theory* (2nd ed., 2003: 173-195). Malden, MA: Blackwell Publishing, Ltd.

Frank, Nancy (2002). Rethinking Planning Theory for a Master's-Level Curriculum. *Journal Of Planning Education and Research* 21:320-330.

Friedmann, John. (1993). Toward a Non-Euclidian Mode of Planning. In Scott Campbell & Susan Fainstein (Eds.). *Readings in Planning Theory* (2nd ed., 2003: 75-80). Malden, MA: Blackwell Publishing, Ltd.

Gärling, Tommy & Kay W. Axhausen (2003). Introduction: Habitual travel choice. *Transportation.* 30: pp. 1-11.

Hatuka, Tali & Alexander D'Hooghe (2009). After Postmodernism: Readdressing the Role of Utopia in Urban Design and Planning. *Places*, 19(2):20-27.

Jacobs, Jane. (1961). *The Death and Life of Great American Cities.* In Scott Campbell & Susan Fainstein (Eds.). *Readings in Planning Theory* (2nd ed., 2003: 61-74). Malden, MA: Blackwell Publishing, Ltd.

LaBarre, Suzanne (2009). Life After Sambo. *Metropolis*, July/August 2009: 52-59.

Lawson, Bryan. (2006). *How Designers Think: The design process demystified* (4th ed.) Oxford: Elsevier, Ltd.

Lennox, John. Pluralism: Do all religions lead to the same goal? *Foundations of Apologetics* (vol. 8) [video]. Oxford, UK: RZIM.

Lewis, C. S. (1980). *Mere Christianity* (Rev. ed.). New York, NY: HarperCollins Publishers.

Lindblom, Charles. (1959). The Science of "Muddling Through." In Scott Campbell & Susan Fainstein (Eds.). *Readings in Planning Theory* (2nd ed., 2003: 196-209). Malden, MA: Blackwell Publishing, Ltd.

Lucy, William K (1988). APA's Ethical Principles Include Simplistic Planning Theories. In Scott Campbell & Susan Fainstein (Eds.). *Readings in Planning Theory* (2nd ed., 2003: 413-417). Malden, MA: Blackwell Publishing, Ltd.

McAllister, Stuart. Truth and Reality. *Foundations of Apologetics* (vol. 2) [video]. Oxford, UK: RZIM.

McGrath, Brain & Jean Gardner (2007). *Cinemetrics: Architectural Drawing Today.* West Sussex, UK: Wiley-Academy.

Nasar, Jack L. & Wolfgang F.E. Preiser (Eds.).(1999). *Direction in Person-Environment Research and Practice.* Brookfield, VT: Ashgate Publishing Co.

Neisser, Ulric (1976). *Cognition and Reality: Principles and Implications of Cognitive Psychology* [out of print]. W.H. Freeman & Co.

Neuman, Michael (1998). Does Planning Need the Plan?. *Journal of the American Planning Association*, 64(2): 208-220.

Perry, David C. (1995). Making Space: Planning as a Mode of Thought. In Scott Campbell & Susan Fainstein (Eds.). *Readings in Planning Theory* (2nd ed., 2003: 142-165). Malden, MA: Blackwell Publishing, Ltd.

Pinker, Steven (2002). *The Blank Slate: The Modern Denial of Human Nature.* New York, NY: Penguin Group.

Scott, James C. (1998). Authoritarian High Modernism. In Scott Campbell & Susan Fainstein (Eds.). *Readings in Planning Theory* (2nd ed., 2003: 125-141). Malden, MA: Blackwell Publishing, Ltd.

Standford d.school (2009). http://www.standford.edu/group/dschol/index.html. Accessed August 11, 2009.

In: Eco-City and Green Community
Editor: Zhenghong Tang

ISBN: 978-1-60876-811-0
© 2010 Nova Science Publishers, Inc.

Chapter 7

TRANSPORTATION FOR GREEN COMMUNITIES: WHAT ARE THE COUNTIES DOING?

Praveen K. Maghelal

Department of Public Administration, University of North Texas, USA

INTRODUCTION

Transportation and Green Communities

The field of transportation and urban planning has been aware and interested in the objectives of creating green communities for about two decades now. Several policies and legislations have been enacted that directly or indirectly recommend transportation and urban planners to deal with the air pollution in their communities.The Clean Air Act (CAA) of 1990 recommended that Metropolitan Planning Organizations (MPOs) should attain the Air Quality standards through network-based models and appropriate land use planning. The Surface Transportation Act of 1991 required the MPOs to plan in consultation with other agencies and indicators to plan efficient travel by all modes. The Intermodal Surface Transportation Efficiency Act (ISTEA) of 1991, the Transportation Equity Act for the 21st Century (TEA-21) of 1998, and the Safe Accountable Flexible Efficient Transportation Equity Act: A Legacy for Users (SAFETEA-LU) of 2005 has build upon the growing need to address the need for a healthy green community. These Acts have focused on accommodating all modes to transportation and reducing the Vehicle Miles Traveled (VMT) by individuals through integrated Land Use Planning.

Although not directly linked to transportation, the issues of deteriorating urban (both environmental and community) health were noticed even before the 1990s. The Clean Air Act of 1955 and its subsequent amendments (CAAAs) have focused on reducing the significant impact of air pollution on the environmental and community health. The Clean Air Act of 1970 actually identified the automobile as a major contributor to the air pollution. Subsequently, the transportation planning agencies were called upon to meet the air quality goals with efficient land use planning.

The Primary reason for this integrated effort was the impact of air pollution on the environment and the individuals of the community. Transportation accounts for 24% of Green House Gas Emissions and 55% of Nitrogen Oxide (NOx) (Bloomberg and Aggarwal, 2008). Reduction in these emissions can have direct impact on public health. Since the air pollutions is directly associated with the respiratory and other chronic diseases, efforts to reduce the air pollution is eminent. For example, the UN Intergovernmental Panel on Climate Change (IPCC) reports that it is important to deal with this issue because, it will "directly shape the health of populations such as education, health care, public health initiatives and infrastructure and economic development" (IPCC Report 4, pg. 48). Several urban and transportation planning approaches have been proposed in the past and recent years to deal with the issue of air pollution

Planning Approaches

Several theoretical and conceptual planning approaches have been discussed for at least a century to organize and commute between different land uses. Ebenezer Howard's *Garden City* (1898) and Frank Lloyd Wright's *Broadacre City* (1935) conceptualized communities with distinctive land uses. The Garden City avoided pollution from production industry to affect the residential communities, thus mitigating the impact of air pollution on public health. However with ever expanding cities, the travel to these distinct land uses require long commute and use of automobile, resulting in increased emission from motor vehicles. F.L. Wright's *Broadacre City* emphasized diversity in unity of land uses. His conceptualization accommodates a car for each citizen of the Broadacre City. Priority was given to building multi-lane highways and little or no public transport or parks were given enough attention.

Frederick Law Olmsted and relatively recent, Peter Calthorpe conceptualized planning approaches that concentrated on the development of park and use of public transportation to deal with the issues of air pollution. Health being at the center of urban issues, Olmsted, in the 1870's, proposed urban parks within walking distances of the residents to improve air quality and increased walking and biking in the park. Similarly, to reduce the travel by private car and increase the usage of public transit, walking, and biking, Peter Calthorpe proposed the pedestrian pockets adjoining the transit locations with varied mixes of land uses.

New Urbanism and Smart growth principles: As recently as 2001, the principles of New Urbanism and Smart growth development have gained immense popularity among planners. The ten principles of New Urbanism seek to develop communities that are actively engaged in walking and biking and mitigate the use of car for work and non-work related travel (CNU). Built environment that can make walking and biking feasible through better connectivity, pleasant architecture and urban design, and accessibility to various destinations are some core principles of New Urbanism. Similarly, the smart growth principles recommend compact development with variety of choices in housing and travel mode (www.smartgrowth.com). In many sense, these principles reflect the attribute of traditional communities that have shown to have lower VMTs compared to suburban or modern neighborhoods (Handy 1996, Cervero 2002, Cervero and Gorham 1995, Cervero and Radisch 1996). Therefore, several planning approaches have been proposed to develop better and greener communities with particular attention to transportation.

Transportation Approaches

Transportation planning has seen as shift in its approach of modeling the travel pattern. Transportation planning has included its focus of the automobile usage, through the Interstate Highway Act of 1956, to incorporate all travel modes. One such approach that has gained popularity, which is reflective of Peter Calthorpe's Pedestrian Pocket, is the Transit-Oriented Development. A study by Loutzenheiser (1997) investigated the pedestrian access to Bay Area Rapid Transit (BART) stations. His investigation reported that car ownership and availability of parking at transit stations were inversely and significantly related to walking to the stations. TODs have several benefits including improved air quality, preservation of land, reduction of sprawl and restriction of urban development along the transit corridors and facilities (TCRP Report. 2004). Another important contribution of the TODs is the development of pedestrian and bike network along the road corridors to support walking and biking to transit station. Beyond the obvious impact of increased ridership and revenue, the report to improve the physical activity levels for the community as a whole.

Increased walking to destinations like transit stations and using transit for both work and non-work related trips has transportation and health benefit through regular physical activity (CDC 1996, Francis 1997, and USDHHS 2000). Studies indicate that neighborhoods with transit destinations within walking distance of the households report higher average walking trips (Handy 1996). A more recent study by Younger et al. (2008) reports that considering health consequences of changes in transportation, building and land use can bring co-benefits in climate change mitigation. Their findings report that "Aspects of the physical environment that facilitate physical activity for all populations offer the co-benefit of reducing motor vehicle associated pollution, thereby diminishing both health hazards and the GHG emissions contributing to climate change".

Current Issues

Efforts to reduce emission: Analysis of the current policies to reduce emission report weakness in current approach to combat the negative effects of transportation on climate (Anable and Shaw 2007). Their study revealed that policies to reduce energy use & emissions from the transportation sector generally fall into three categories: a) promoting technical advances in fuel carbon coefficients & engine efficiency; b) targeting modal switch; and c) attempting to reduce distance traveled. However, less emphasis is places on the last 2 categories. Similarly, a study by Fergusson and Skinner (1999) reported that to reduce the adverse effects of transportation on climate and ecosystem, research to understand the ecological effects of transportation needs to be supported. Therefore, understanding of the current issues, and current policies, barriers to implementation of those policies, and the opportunities to deal with these barriers and issues needs to be identified.

ANALYSIS

Efforts to reduce emission has always, at least till recent, was a top-down approach, where local governments were regulated to meet standards set by federal agencies in the US. However, studies report that for effective reduction, the local agencies have to be proactive in developing policies subjective to each region. The MPOs can play an important role in coordinating this effort at regional level, especially because MPOs are responsible to develop the Long Range Transportation Plan (LRTP) in coordination with local agencies. The LRTP have to address the issues of Air Quality Control and propose developments that can help relieve the traffic congestion in the region. The LRTP includes strategies, policies, and recommendations that local agencies can use to improve the transportation connectivity and accessibility in the region. Therefore, this study analyzed the current and future efforts in the eight counties of the South Florida region and also inquired the awareness, efforts and barriers as described by the decision making stakeholders from each of these counties. The major objective of this study is to analyze the opportunities and barriers as perceived by decision makers in transportation and planning agencies.

The focus areas of this study were the counties that have a MPO and develop a LRTP every five years to address the issues and developments related to transportation. Eight counties: Charlotte, Lee, Collier, Miami-Dade, Broward, Palm Beach, Martin and St. Lucie in the South Florida region were identified for this study (Table 7a and Figure 7a). About 75% of individuals who commute to work in these counties drive alone. The share of workers who commute for 25 minutes or more every day, vary across the counties. Charlotte reports the least with 31.5% of its commuter travel time being 25 minutes or higher while Miami-Dade reports a higher share of 54%. This indicates that at least 46% of individuals who travel to work within 25 minutes can be targeted to make a shift in the travel mode and use public transit, walking, or biking for their commute. Since the share of vehicular emission is high in all but the Martin County, it is imperative to take measures to reduce the emission from the vehicles. This can help to reduce the air pollution in these counties and thus reduce the impact on environmental and community health.

Table 7a. Transportation and Emission Characteristics of the study counties

	County Name	Drove Alone	> or = 25 min travel time to work	Percent of On Road Emission*	Total Million Metric Tons of Carbon Emission*
1	Broward	80%	48.90%	43.07%	5.063815039
2	Charlotte	81.70%	31.50%	68.56%	0.379358273
3	Collier	74.40%	37%	62.15%	0.699229842
4	Lee	78.70%	39.90%	41.76%	1.901191711
5	Martin	79.20%	38.60%	10.42%	2.761734297
6	Miami-Dade	73.80%	54%	58.14%	5.068495052
7	Palm Beach	79.60%	41.80%	51.81%	2.917015697
8	St. Lucie	80%	43%	63.15%	0.544399968

* Source: Vulcan Data.

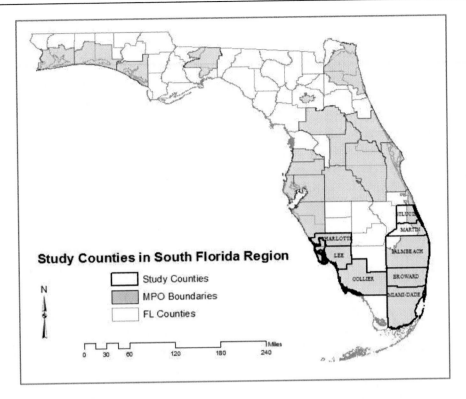

Figure 7a. Eight Study Counties of South Florida.

Stakeholder participation is critical to success of plan creation and implementation to deal with air quality control. It is therefore important to acknowledge the understanding of decision-making stakeholders in creation of such plans. Subsequently, the understanding and awareness of transportation and urban planners about the issues of transportation and emission can help determine the barriers and opportunities for improvement of the LRTP. Therefore, the decision-making stakeholders such as chief planners, transportation and urban planners, and commissioners were inquired regarding their awareness, actions and future plans to deal with the emission issues in their respective county. Three core semi-structured questions were used as guideline for this inquiry:

1) Are the stakeholders aware of the issues regarding climate change in their county?
2) How are they addressing the existing issue? What are their specific actions?
3) What are the actions that they plan to take in future to address the climate change issue?

During discussions, the barriers to deal with current and future emission issues were inquired. Both political and administrative barriers were inquired and discussed. They were also encouraged to discuss opportunities as they saw fit to deal with the transportation emission and green communities in their county.

IMPLICATIONS TO CREATE GREEN COMMUNITIES

Decision-making stakeholders were requested to volunteer to interview through phone-call and emails for each county. The pool of interviewees included the senior planners, commissioners, transportation planners directly involved with LRTPs, and a MPO Director. In all, 12 stakeholders were interviewed using the semi-structured interviews. Informed verbal consent was obtained from the participants and were given a brief about the purpose of the interview. Issues of emission and its impact on climate change were discussed as part of the interview. Participants were inquired about their awareness, action, and future plans to deal with the issues of emission from auto and climate change.

Stakeholder Awareness

The semi-structured qualitative interview identified two distinct groups of counties. Counties those are passionate and committed to issues of emission and climate change and others who are not aware of any specific issues in their county (St. Lucie and Hendry). Counties that were aware about the issues of emission and climate change reported specific issues that needed immediate attention. Some of issues that were repeatedly mentioned were:

- Automobile exhaust
- Sea level rise and its consequential impact to the County and region
- Horrible hurricane
- Erosion
- No or sparse rain
- Sprawl

While the issues of emission and climate change were almost common, certain counties were more committed to deal with the current issues and had taken actions to do so. For example, Martin County has identified the policies in response to the current issues and created an Evaluation and Appraisal Report which outlined these policies. The Miami-Dade County has an active Climate Change Advisory Task Force to deal with the current issues of transportation and emission control. Strategies to reduce emission were proposed by the decision making stakeholders included:

- Improve the flow of traffic to reduce idle travel time,
- LED for street lighting
- Fixed bus routes improving the transportation routes

Acts in Progress

Counties that have proactively dealt with the climate change, and particularly the emission issues, have proposed actions to improve the air quality and mitigate emissions from auto. These actions include better land use planning, greener developments, and availability of alternative mode of travel. Specific actions include:

- Coordinating the transportation and land use planning
- Development of plans/projects in the pipeline that are hoped to discourage single occupancy auto travel, i.e. promoting transit oriented development
- Discouraging density away from non-transit corridors
- Establishment of a new/improved public transportation system
- Not have increased number of single family neighborhoods
- Alleviate future develop impact on the environment
- Interagency coordination (planning & transportation) to find ways to reduce carbon emission
- Encourage re-use of underutilized developed areas
- Restricted urban development inside an Urban Service District resulting in 75 percent of the County to remain available for Agricultural production
- Encouraging energy efficient construction
- County acquisition of land for conservation
- Creation of a solar energy farm
- Preserving rural and agricultural lands
- Implementing four-day work weeks

Although the specific strategies, actions, and policies as identified by the counties can help reduce the impact of emission on environment and individuals, stakeholders face barriers that either hinders the implementation or continuation of their intended actions. The most common barrier across most counties was the availability of funding to carry out their proposed actions. Although funding from the state and federal agencies results in identification of strategies and actions, the implementation requires larger funding mechanism from these agencies. The specific barriers as indicated by the stakeholders were as listed below.

- Increase in population
- Costs of finding viable alternative energy source
- Costs to revamp transportation infrastructure
- Lack of general public understanding of the magnitude/seriousness of the issue
- Political support and affiliation
- Power struggle
- Financial support
- Irregular zoning

Next Step to Create Green Communities

All counties, irrespective of their level of current motivation to deal with emission issues, identified actions that they plan to integrate in their 2035 Long Range Transportation Plans.

These included strategies to increase intergovernmental coordination, rezoning and improved land use and transportation plans, and providing infrastructure for alternative modes of travel. The specific recommendations included:

- Higher level of local political will, consensus, and commitment
- Change in transportation/transit policies at the local level
- Implement employee fee for parking
- Discount transit fee
- Public consensus to determine if climate change issue is important enough for new legislation to be implemented
- Look holistically at policies relating to land use, transportation, building codes, landscaping standards
- Re-zoning and New Land Use Policies
- Continue to support bicycling and walking as viable means of transportation
- Public outreach and education
- Private-public partnerships to address climate change issues
- Locating bio-diesel or solar plants within the county (Hendry)
- Implementation of gas tax
- Use of Alternative fuels such as Biodiesel and ethanol.
- Expand the hybrid vehicle fleet
- Land use alternative scenarios which support multi- model travel

Transportation Planning for Green Communities

Four major themes emerged from the discussion with the participants of the interview from the eight counties in South Florida region: Educational; Institutional; Political; and Financial (Figure 7b). These themes address the core issue of vehicular emission reduction strategies by understanding the current issues and barriers and identifying the opportunities to create a greener transportation.

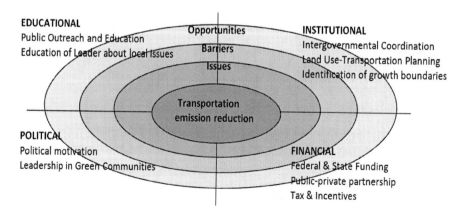

Figure 7b. Barriers and Opportunities for Transportation Emission Reduction.

Educational Issues, Barriers, and Opportunities

This theme indicates educating both ends of the public spectrum (decision-makers and citizens). The decision makers and leaders need to 'Think Globally, Act Locally'. They need to educate themselves with the recommendations by the IPCC but also understand its feasibility and pragmatism of implementing such recommendations in the local communities. Local resources and interest need to be kept in mind when making decisions to adapt the recommendation. Also included in their understanding is the need to educate the local public about the objectives and benefits of using greener transportation, both locally and globally. Increased public outreach and education is important to the successful adaptation of the recommendation by the counties.

Institutional Issues, Barriers, and Opportunities

As decision-makers, all governmental agencies need to coordinate their efforts better in understanding and addressing the issue of emission reduction. Important to their effort is the intergovernmental coordination across agencies at local, regional, and federal levels. Policies and recommendation across the different levels of government should be carefully drafted for proper coordination of action and responsibilities. At the regional and local level, the comprehensive land use plans and the long range transportation plans should be coordinated even more to propose developments that support greener modes of transportation. Also, developments within the county should adhere to zoning regulation, building construction techniques, and transportation infrastructure that do not infringe upon the agricultural and rural land. Therefore, growth boundaries need to be identified and created to restrict the sprawl of development and hence reduce the travel time and needs to use auto to travel to and from work.

Financial Issues, Barriers, and Opportunities

One of the most important facilitator of green communities is the availability of funds to go green. All the participating counties reported that although technical and strategic support is available from the state and the federal agencies, the limitation of funds restricts their capabilities to provide infrastructure and support needed at local level. Availability of funds from these agencies can help the local agencies kick start the development that can help reduce the emission on road. Another strategy is to develop a public-private partnership so that private agencies can provide necessary infrastructure such as to set up a bio-diesel industry or windmill to help move to greener and healthier communities. Local and state agencies can contribute to these industries through tax breaks and other incentives that can help them keep their production cost low. Also, the state and local agencies should provide incentives to individuals who use and promote the use of alternative modes of travel.

Political Issues, Barriers, and Opportunities

Party affiliation and individual motivation of leaders can benefit or hinder the actions to reduce emission in the counties. Political leaders, irrespective of their affiliation, should try to understand and meet the demand for green approaches that can benefit their constituents and make them pioneers of green communities. Priority, now rather than later, should be given to policies and regulations that can support development of green infrastructure at local level. Leaders with motivation to create green communities can put forth legislature to support green infrastructure, draft proposals for funding, and push elected officials and local agencies to take actions to make travel more environmental and public health friendly.

ACKNOWLEDGEMENT

The participants of the interview from the eight counties in the South Florida region provided valuable inputs that helped in understanding the role of county agencies. Graduate students enrolled during the Spring 2009 in the Introduction to Transportation course in the School of Urban and Regional Planning at the Florida Atlantic University helped conduct the interview and their help is greatly appreciated.

REFERENCES

Anable, J., and Shaw, J. (2007). *Priorities, policies and (time) scales: the delivery of emissions reductions in the UK transport sector.* Royal Geographical Society (with The Institute of British Geographers). Vol. 39 No. 4, pp. 443–457

Bloomberg, M.R., and Aggarwala, R.T. (2008). Think Locally, Act Globally. How Curbing Global Warming Emissions can Improve Local Public Health. *American Journal of Preventive Medicine,* Vol. 35: No. 5, pp. 414-23.

Cervero, R. (2002). *Built environments and mode choice: Towards a normative framework.* TR: D 265-84

Cervero R, Gorham R. (1995). Commuting in transit versus automobile neighborhoods. *J Amn Plan Assoc* 61: 210-25

Cervero R, Radisch C. (1996). Travel choices in pedestrian versus automobile oriented neighborhoods. *Trans Polic* 3(3):127-141.

Cervero, R. et al. (2004). *Transit Oriented Development in America: Experiences, Challenges, and Prospects.* Washington, D.C.: Transit Cooperative Research Program, Report, 102, 2004.

Cervero, R. (2007). Transit Oriented Development's Ridership Bonus: A Product of Self Selection and Public Policies, *Environment and Planning A*

Cervero, R., Dunphy et al. (2005). *Development Around Transit.* Washington: Urban Land Press

Climate Change (2007): Synthesis Report. 2007. *An assessment of the Intergovernmental Panel on Climate Change* (IPCC). UN Report

Congress of the New Urbansim accessed at http://www.cnu.org/ on August 25, 2009.

Fergusson, M., and Skinner, I. (1999). Greening Transportation Toward a Sustainable Future: *Addressing the Long-Term Effects of Motor Vehicle Transportation on Climate and Ecology. Environment*. Vol. 41.1, p. 24(4).

Handy, S. (1996). Urban form and pedestrian choice: study of Austin neighborhoods. *Trans Research Rec* 1552: 135-44.

Hanson S. and Giuliano, G. (2004). *The Geography of Urban Transportation*. The Guildford Press. New York, NY.

Holtzclaw J. (1994). *Residential patterns and transit, auto dependence, and costs*. San Francisco: Natural Resources Defense Council.

Kitamura R, Mokhtarian PL, Laidet L. (1997). A micro-analysis of land use and travel in five neighborhoods in the San Francisco Bay area. *Trans* 24: 125-58.

LeGates, R.T. and Stout, F. (1999). *The City Reader.Routledge: Taylor and Francis Group*. New York. NY.

Loutzenheiser D. R. (1997). Pedestrian access to transit: Model of walk trips and their design and urban form determinants around bay area rapid transit stations. *Trans Research Rec* 1604: 40-49.

Smart Growth accessed at http://www.smartgrowth.org/Default.asp?res=1680 on August 25, 2009.

Younger, M., Morrow-Almeida, H. R., Vindigni, S. M., and Dannenberg, A. L. (2008). The built environment, climate change, and health: Opportunities for co-benefits. *American Journal of Preventive Medicine,* Vol 35: 517-526.

SECTION IV: MODELS OF ECO-CITY AND GREEN COMMUNITY

In: Eco-City and Green Community
Editor: Zhenghong Tang

ISBN: 978-1-60876-811-0
© 2010 Nova Science Publishers, Inc.

Chapter 8

GREEN URBAN PATTERN AND ENVIRONMENTAL DESIGN IN HONG KONG

Shaojing Tian

Community and Regional Planning Program
University of Nebraska, Lincoln, NE, USA

INTRODUCTION

More than half of the population in the world lives in urban area. Urban environment has experienced a rapid growth and influx of population, and borne the major function of the operation of society and economy. We witnessed the rapid development of East Asian cities including Hong Kong, Tokyo, and Shanghai. Among them, Hong Kong is the most iconic vertical city in the world with attractive ridgeline and collective skyline of skyscrapers as symbols of modern urbanism. As a harbor city, Hong Kong has grown from a colony trading port more than a hundred years ago to the world's leading financial and commercial center. Always being compared with New York City, Hong Kong is viewed as the place where many of the visionary ideas proposed by 1920's New York architects came true. Even today, Hong Kong has surpassed New York in terms of the number of high-rises, population density, and efficiency of public transit services. Besides the accomplishment it achieved in economic and social growth, as the most densely populated and concentrated city in the world, its urban form, land use pattern, and impact to the coastal ecological environment with its unique location merit further study and review on it.

OVERVIEW OF THE HONG KONG'S URBAN PATTERN

Urban Pattern Development in a Historic Context

Urban population is now generally increasing at a speed three times faster than overall population. As one of the most densely populated places in the world, Hong Kong faces the challenge of accommodating 7 million people in this 428 square miles hilly island, and 30,000 persons per square miles in its most dense area. (Sang and Chan, Building Hong Kong: Environmental considerations, 2000, P4) Hong Kong Island, Kowloon and New Kowloon were the old urban areas. According to 1976 By-census, they had their densities of 13,192, 83,104, and 39,729 per square km respectively. They were 3 to 20 times the overall density in the colony time. (Fung-shuen, Urban Hong Kong, 1981, P14). With growth of the native-born Hong Kong population and the influx of immigrant from China mainland during the time of Chinese Civil War, the population of Hong Kong boomed at a rapid pace, and several areas had experienced fast development. These areas have been becoming new towns in the later decades, including Tsuen Wan, Tuen Mun, Sha Tin, Tai Po, Fanling-Sheung Shui and Yuen Long. Tsuen Wan, Sha Tin and Tuen Mun, had experienced significant population growth during 1960 to 1980s. By 1970s, Tsuen Wan had developed into part of the metropolitan area, with its density comparable to Hong Kong Island. Except for the areas which developing into new towns, the density of the rest rural parts of the New Territories remained low. (Fung-shuen, Urban Hong Kong, 1981, P14).The highest density areas are in the older urban area of Kowloon peninsula, with a population of 2 million and density of 111,500 per square miles according to 2006 census. Together with Hong Kong Island, Kowloon contains almost half of Hong Kong's total population.

Urban Pattern with the Terrain Context

When we examine the spatial urban form in such a hilly island at the unique location, we can find that the land use pattern of Hong Kong is formed by the development based on its topography and terrain. Like most of the other port cities, the old tradition and mature urban areas are close to the waterfront where the city started. As the population grew and economic sectors diversified, the city developed with coastal occupation and building upward. The most urbanized area with the financial and commercial center developed along a seaward sprawling corridor of Hong Kong Island and Kowloon peninsula. The famous urbanism scenery with tall skyscrapers in the CBD, Hong Kong Convention and Exhibition Center, Central Government Office, HSBC Headquarter Building are all at the north of the Hong Kong Island. Crossing the beautiful Victoria Harbor, the other side is the south of the Kowloon peninsula with Central District of Tsim Sha Tsui. The urban area in Hong Kong Island and Kowloon Peninsula is bounded by the steep hills of high elevation mountainous area in the central of both islands. The hilly topography and irregular shape exercised controls over the urban spatial pattern of Hong Kong development.

With its small size, highly irregular shape and its independent location, Hong Kong survive, prosper and even grow into one of world's leading cities. Hong Kong is promoted and enabled to be successful by its efficient transportation network and the ferry services across harbor which link the Hong Kong and Kowloon together. This point can be regarded as an aspect of the constructing successful urban structure which fulfills the need of such a fast metabolic city. Another apparent aspect of Hong Kong's successful urban structure comes from its advantage of vertical terrain. The high-class residential districts are separated from other districts by the choice of sites according to their elevation on mountain. In Hong Kong Island and Kowloon, the high-class residential zone is determined by elevations, so that the vertical dimension positively correlates with social and economic status. (Fung-shuen, Urban

Hong Kong, 1981, P41)The lack of available sprawl space and short distance from the harbor front to the steep hills make Hong Kong has the world's greatest number of skyscrapers. As the most vertical city, Hong Kong has most of its population living or working above the 14th floor than anywhere else on Earth. Hong Kong's vertical advantage has also being applied to its efficient transportation infrastructure (underpass, overpass, and elevated walkways), and open space network with its outstanding vertical landscape design (green roof and podium garden). This vertical urban pattern enable such a high-value land to survive and prosper, meanwhile, creating a diversity of vertical landscape view and make it successfully exemplify the conception of vertical urbanism.

Urban Pattern as Compact City

The tall building as the principle instrument of metropolis urban design reflects both the high land values and the development pressures in Hong Kong. The compact cluster use of commercial, residential and industrial district with tall buildings, and the high efficiency of transportation make Hong Kong successfully fulfill its needs of fast metabolism. The provision of infrastructure, public transportation, housing and jobs would be easy to be served by the economy in a small scale. More than half of the Hong Kong's urban population has being concentrated to the new towns, and the citizens are housed within easy commuting distance of the urban area. Most of the densely populated residential districts in Hong Kong are located near the efficient public transportation lines and intersections. The residents take the advantages of the public transit services and conveniently travel from home to work and to other urban retail, shopping and entertainment areas. Due to the full utilization of the public transit services, citizens in Hong Kong are less reliant on personal vehicles, which result in the lower car ownership compared with other developed countries. Another advantage of the high density is slowing down the speed of urban sprawl and helping conserve natural and agricultural land. As a compact city, Hong Kong also has fewer energy problems and manages to conserve the energy in terms of lower energy consumed per capita. As the new urbanism and smart growth planner in United States have realized the need to reshape the urban pattern to a relatively increased density compact form in order to reduce to speed of urban sprawl, Hong Kong offers a good example of such urban patterns. In the high density residential, commercial and office district, large amount of planting area in vertical landscape has being introduced to everywhere. Other than the public and community park, the green roof and podium garden which are widely used in the high density areas, provide public visual accessibility to the open space and beautiful streetscape. Citizens can visually enjoy the planting landscape or get involved in the activities in the nearby parks conveniently without travelled a long distance from home. At the same time, Hong Kong has also thrived on its mix of uses and diversity, which establish a self-sustained urban form for a multiplicity and pluralism of urban uses. One of the scenarios that Hong Kong is well known for by tourists is the enjoying the dinner with friends in a little restaurant after spending a whole afternoon on shopping in small vendors. The old urban areas constitute part of the life and culture in Hong Kong, providing a robust and compatible mixed-use of workplaces, retail, entertainment and restaurant. This brings the connectedness and interaction of people and builds up the urban culture in Hong Kong, moreover, helps consolidate the traditional economic and social development of the city.

PLANNING WITH ISSUES ON THE ENVIRONMENT AND ENERGY

As the old saying goes, beauty is in the eye of the beholder. Tourists are always fascinated by the attractive ridgeline, collective skyline of modern skyscrapers, and beautiful Victoria Harbor from the cruise ship. However, from the eye of some citizens, especially those in the old slums, the beautiful open space would possibly be nothing but illusions. (Michelle Huang, Walking Between Slums and Skyscrapers: Illusions of Open Space in Hong Kong, Tokyo, and Shanghai.) From the planner's eye, on the "blue and green print", the blue represents the Victoria Ocean; the green represents the island with mountain and open space, the grey represents the build-up area with infrastructure. We have already seen the grey's encroachment of the green and blue area. Do we have to occupy more of the rest the green to accommodate the increasing population and development?

Urban form has long being considered as a dynamic process with the physical, economic, social, and civilized changes. Human kind and ecological issue are considered to be the basis of urban planning. What kind of urban pattern we are going to build in urban renewal? Can we build an urban form with highly-functioning and efficient social and economic operation without compromising the future generation's ability of using the ecological environment and natural resources? These are questions contemporary urban planners keep contemplating. Located on the hilly island and the boundary of ocean and hinterland, Hong Kong's urban pattern posts a considerable impact to the coastal wetland ecology and a climate change to the hinterland. In the following part, the paper will further discuss the urban pattern and the future of sustainable urban renewal of Hong Kong with the major consideration of ecological environment, energy, and human health.

Open Space – Quality and Quantity

With its unique island location atop the mountain as well as the subtropical warm and humid climate, Hong Kong has made achievement on the construction of green space with public and community parks, streetscape plantings, and elevated podium gardens which have created a pleasant visual landscape to most part of city. However, environmental issue should be taken into account in urban open space design rather than considering only for "function" design and good aesthetics. It bears the responsibility of improve the urban ecological environment and change the microclimate of urban area which is considered as the heat island.

Hong Kong's green infrastructure can be divided into three categories, with ornamental gardens in the center, extensive green space in the inner city, and the agricultural farm in the outer city. However, the three kinds of green unites are isolated with each other due to the lack of robust connect of green belts. The lack of connectivity and accessibility results in the fragmentation of urban green space. The fragmental urban ecological environment and extremely homogenous biodiversity make it hard to fulfill its needs from the ecological perspective. An effective way to improve the urban ecological environment is to incorporate the "nature jungle" into the urban impervious surface. Besides an adequate system of open space, a well-functioning green infrastructure should form an integrated network to connect the parks, open space and green belts, and eventually tied them back to the mountain areas in Hong Kong.

Another weakness of Hong Kong's green infrastructure lies in the waterfront corridor in the south part of Kowloon Peninsula and north part of Hong Kong Island which play a critical role to the coastal wetland ecology and the climate change to the hinterland. Most of the waterfront zones are dominated by the vehicle-oriented transportation belts, with fewer pedestrian friendly environment and green space. To reduce the adverse impact of urban construction to the coastal ecology and climate, waterfront areas are worthy of emphasis on more green space with pedestrian-oriented systems. Continuous pedestrian and bicycle path should be incorporated in the green corridors and severed as the linkage of an adequate framework of green space with parks, green roof, and podium garden. While providing complementary improvement to the urban green space, it also creates easy access for public and directs them to the exercise and job recreation facilities in the open space.

The remarkable achievement of Hong Kong is the aesthetic design to create attractive urban landscape and streetscape. However, the quantity of open space is too little to be well-functioning in terms of ecological and environmental concern. Hong Kong has more than half of its total territory covered by green land, but almost all of them are in higher mountain area or outskirt countryside. From an overall picture, most of the developed urban area is occupied by impervious surface, concrete building and infrastructure with limited greenery landscaping among them, especially in the seaward commercial and industrialized area which was clustered by CBD buildings. Even the countryside is encroached of vegetation and wetland for development of more housing, employment, industry, transportation network and other infrastructure. Early urban development in last century intended to give way for population growth and economic development, but paid little attention to environment. The existing buildings layout and spatial limitation in metro area makes the greenery landscaping and construction more difficult to achieve an integrated level. Improving existing open space in compact district and conserving more rural land in future development are effective ways to compensate the value of the natural land and vegetation that we have lost. In the most vulnerable, high-density area without ecological safeguard, such as Sheung Wan, Central, Wan Chai and Causeway Bay, replace bald impervious surfaces and non-occupied areas (ground street, roof, podium) with greenery as many as possible if conditionally permitted. Incorporate more trees and shrubs to pedestrians and link them with adjoining old districts and promenades. Introduce green space and landscaping to the newly built residential and industrial community in New Town and New Territories. According to the Hong Kong Planning Standard and Guideline, standards for provision of open space (square meters/person) are set in urban areas, public housing estates and comprehensive residential developments, industrial-office, business, commercial areas, and rural areas. These standards can be used as minimum requirements of open space quantity.

Spatial Form and Tall Buildings Design – Wind, Sun, and Plants

Hong Kong is famous for its various architectural types and shapes of skyscrapers. Tall buildings in Hong Kong constituting the collective image of skyline are principle instruments of vertical urbanism and symbols of a strong sense of local identity. While Hong Kong's high urban density and efficient public transport use make average energy consumption per capita relatively low, around 50% of all energy use is associated with tall buildings. (Sang and Chan, Building Hong Kong: Environmental considerations, 2000, P26). The distribution of energy consumption in building mainly lies in mechanical and electrical systems. Hong Kong's microclimate, in particular, calls for the energy savings in mechanical cooling and artificial lighting. (Sang and Chan, Building Hong Kong: Environmental considerations, 2000, P27). Therefore, light, wind, air, and solar heat are environmental factors to be considered in strategic urban design. Because of the high-density in urban area and the subtropical climate with hot and humid summer, Hong Kong suffers from the effect of heat island. To improve microclimate condition in urban area thus reduce the energy consumption used for cooling system, building design needs to be associated with environmentally friendly and low-energy concern.

Air ventilation is one the solutions to use winds for thermal relief and maintain human comfort. Breezeways along the prevailing wind direction should be able to penetrate through the urban building area to promote the air movement and circulation in order to remove heat and humidity. (Hong Kong Planning Standards and Guidelines. December 2008, From http://www.pland.gov.hk/tech_doc/hkpsg/english/index.htm). Tall structures form urban canyons which trap the air circulation and concentrate the air and noise pollution. Narrow urban canyons in old districts like Mong Kok are planning to be reshaped to widen the setbacks on each side of the streets in urban renewal. Hierarchy building setbacks and spatial relationship between low-rise and high-rise buildings are taken into consideration for air ventilation.

"The array of main avenues where buildings are located on each side should be aligned in parallel or with small angle to the prevailing wind direction, to maximize the penetration of wind through the district. Waterfront is the gateway of cool sea breezes, so locate low-rise buildings rather than high-rise buildings at waterfront area to avoid blocking the winds. ...... In general, gradation of building heights would help air circulation. Increase the heights towards to direction where the prevailing wind pass through would promote air movement. " (Hong Kong Planning Standards and Guidelines. December 2008, From http://www.pland.gov.hk/ tech_doc/hkpsg /english/index.htm).

Besides air ventilation, reduction of solar heat gain in the summer is another effective way to conserve energy used in tall buildings. This requires involvement in the analysis of sites, which relates to the direction of sunshine and shadow effects. Shading devices like fins, overhangs, and balconies can be used to cut the sunlight. In Kam-sing Wong's Use of Technology to Assist Environmental Design, (Wong Wah Sang and Edwin H.W. Chan, Building Hong Kong: Environmental considerations, 2000) several technological solar-shading devices for heat relief in tall buildings were demonstrated. Cantilevered roof which extend from the edge of wall can be used for solar collection while providing supplementary solar-shading and insulation. External shading screen wall and bay window are effective ways to reduce the solar radiation to a certain degree. In addition, sophisticated glass materials can also be used to reflect and block extra sunlight. Technologies and materials which can be used to absorb solar radiation and convert to other forms of energy in buildings really merit more study and research in future.

Besides engineering methods, biological solution can also help environment improvement and energy saving in buildings. Vegetation and planting are effective ways to cool down building temperature, reduce radiation gain of buildings, and filter polluted air. Exterior wood plant canopy in ground level provide shade to the low stories interior spaces.

Planting in mid-level such as green roof and podium garden helps reduce the surface temperature of roof and façade of building by process of plant's evaporation. As plants help process carbon dioxide and generate oxygen to clarify air and airborne pollutants, they have being introduced to interior space such as atrium and sky walkways between wings of buildings. In addition to providing green transition areas between internal working areas and the outside, the planting and soft landscape areas also serve as network to connect green space in an elevated level.

Ecological Conservation – Coastline and Wetland

Hong Kong's fast GDP development is coming along with the deterioration of its ecological environment and natural resources. The rural and coastal land has experienced encroachment of vegetation and wetland for the development of housing, employment, industry and urban infrastructure. The continuously increased population in recent years makes its local natural environment worse than ever before. Natural landscapes in Hong Kong are precious habitats for many characteristic animals and plant communities. The reduction and destruction of them may disturb the balance of natural ecosystem and wild life, thus threaten the urban sustainability. In order to protect and maintain the biodiversity with the concern of urban sustainability, natural landscapes and habitats should be conserved with social and economic considerations.

According to the town planning ordinance in Hong Kong, natural landscapes are divided into Country Parks (and Marine Park), Coastal Protection Area, Green Belts, and Sites of Special Scientific Interest.

As mentioned earlier, the coastal ecosystem is unique resources for Hong Kong, but unfortunately, it has being compromised by its early urban design which mainly focused on economic development. The coast front area in such a harbor-city is almost inaccessible for public use, because major part of it is occupied by traffic and industry oriented uses such as sea freight shipping, highways, wholesale markets and sewage treatment. The high density of build-up areas give little space and flexibility for aggressive and innovative design for waterfront in Hong Kong. Even the old beautiful harbor scenery with natural mountain peak and ridgeline is being compromised by the overwhelmed surge of tall concrete structures. (Ted Pryor and Peter Cookson Smith, What Kind of Harbor City Do We Want? Sang and Chan, Building Hong Kong: Environmental considerations, 2000, P63) We cannot help contemplating what role new urbanism can play in Hong Kong to renew the coastline environment in a more integrated and comprehensive process. While advocating the prohibition of excessive urban development for protection from coastal ecosystem erosion, can we venture to consider of revitalizing it for accessible public use and tourist attractions with economically viable use of land of private investment? The coastline is precious resources of Hong Kong people and deserves appreciation, both from the perspective of joyful visual experiences in a beautiful and natural harbor, and from the coastal environmental conservation concern. The coastline renewal in Hong Kong calls for consideration with integrated elements of revitalization, ecological restoration and economic regeneration in a longer term.

There are approximately 3,700 acres wetlands around Mai Po Marshes and Inner Deep Bay area in the northwestern corner of Hong Kong, and have being known as a haven for migratory birds for decades (Hong Kong Planning Standards and Guidelines December 2008). This wetland is an intertidal area with dwarf mangroves, shrimp ponds and fishponds and provides a transition between dry land and water bodies. As an important part of ecological environment near urban area, wetlands have a valuable flood control function, and are also very effective at filtering and cleaning water as to help reduce and eliminate water pollutions. Being considered to be the most biologically diverse of all ecosystems, wetlands are precious habitats for wildlife species, and play a critical role in protection ecological biodiversity and maintenance of urban sustainability. With rapid population growth in Hong Kong, parts of the wetlands were exploited or destroyed mainly for residential uses. With increasing awareness and concern of the ecological role that wetlands play, policies and ordinances were enacted for the conservation and wise use of wetlands since last decades. The wetlands in Mai Po and Inner Deep Bay area have being designated as a Ramsar Site under the Convention on Wetlands of International Importance. To protect the most important wetland resources in Hong Kong, the core part of Mai Po Marshes, as well as the intertidal mudflat and mangroves areas, the inner Deep Bay is listed as Restricted Area under Protection Ordinance. The Town Planning Board has also designated special land use zones and issued guidelines for development within Deep Bay area to reduce adverse impact on the wetland.

Environmental Protection – EIA

As a fast metabolic metropolitan city, Hong Kong is faced with pressing problems of pollution including air and water pollutant, noise, and ecosystem decay. Anti-pollution strategic design is imperative to be incorporated in urban planning for Hong Kong. In order to reduce or eliminate the impact of development actions and the pollutant generated by them. The Environmental Impact Assessment (EIA) Ordinance has being enacted to assess the environmental impact of certain project proposals. The system demonstrates criteria and guidelines for assessment of air quality, water pollution, noise, ecology, water management, and sites of cultural heritages in quantitative terms. Projects such as Roads, railways, airports and sewage system are Designated Projects, which require an environmental permit from the Environmental Protection Department (EPD) before the projects commence. Matters that specified in an environmental permit include: the design, layout of a designated project, physical scale and scope, quality and quantity of discharge and emission of pollutants, mitigation of the environmental impact, pollution control and environmental protection measures. Not only does the system assess the potential environmental impact of development project, but also propose and evaluate efficacious means to minimize the impacts. The application of EIA to Hong Kong urban planning indicates the steps moving towards to an environmental and sustainable integrated planning era.

CONCLUSION

As one of most iconic metropolitan cities, Hong Kong's achievement in social and economic development has being witnessed by the world. With its early role as a trading port, the urban development pattern was determined by its various business and commercial functions. Its unique location of small size and independent islands, and the highly irregular topography make Hong Kong's land very valuable and also bring great pressure to development, which finally created a prosperous compact city with an efficient and well-functioning urban pattern. However, the successful social and economic development is coming with the compromise of natural environment and the aggravation of urban ecosystem. Overwhelming concrete structure, limited open space, homogenous biodiversity, increasing energy consumption and pollution make urban environment deteriorate in a long-term. Human kind and urban ecology are the basis for planning. To create an integrated and sustainable urban ecosystem, the interdependence of people and nature should be taken into account in a balanced ecological planning process. Urban renewal in Hong Kong should be achieved by incremental improvement with the vision combined with physical, economic, environmental and ecological concern.

REFERENCE

Bristow, R. (1984). *Land-use Planning in Hong Kong: History, Polices and Procedures*, Oxford, New York: Oxford University Press.

Gomez, F., and Jabaloyes, J. (2004). *Green Zones in the Future of Urban Planning*, (ASCE)0733-9488(2004)130:2(94).

Hong Kong Planning Standards and Guidelines. (HKPSG, as at December 2008). From http://www.pland.gov.hk/tech_doc/hkpsg/english/index.htm

Huang, T.M. (2004). *Walking Between Slums and Skyscrapers: Illusions of Open Space in Hong Kong, Tokyo, and Shanghai*, Hong Kong: Hong Kong University Press.

Ng, E. (2005). Passive and Low Energy Cooling for the Built Environment, May 2005, *Towards Better Building and Urban Design in Hong Kong*, antorini, Greece.

Sang, W, and Chan, E. (2000). *Building Hong Kong*, Hong Kong, Hong Kong University Press.

Sit, V. F. (1981). Urban Hong Kong. Summerson Eastern Publishers Ltd.

In: Eco-City and Green Community
Editor: Zhenghong Tang

ISBN: 978-1-60876-811-0
© 2010 Nova Science Publishers, Inc.

Chapter 9

GREEN AGENDA IN INDIAN CONTEXT: REFLECTIONS FROM ECO-CITY EXPERIENCES IN PURI, INDIA

Akhilesh Surjan[*] *and Prasanta Kumar Mohapatra*[†]
United Nations University, Japan

ABSTRACT

Urban growth in India is complimented by both perils and promises. 'Eco-City' and 'Green Community' largely remains alien concepts for Indian cities; however, there are some pilot efforts to steer attention on these notions. This paper begins with reviewing Indian policies and plans towards urban development and environmental improvement which may be considered as precursor to Eco-City ideas. Further, field realities from the city named Puri, which was selected to develop as Eco-City in the beginning of this century, have been summarized. A brief qualitative analysis of key issues and concerns of Puri city reflects that it stands as an ideal candidate city to be developed as Eco-City. The paper concludes that, institutional players although played important role to impose Eco-City principles while addressing Puri's concerns, but lacked long term vision and necessary support mechanism to translate Eco-City into a reality with earlier envisioned multiplier effect.

PROLOGUE

India was and is still primarily rural. "The percentage of Urban Population in India has risen from 17.97 per cent in 1961 to 27.78 per cent in 2001. Although 2 out of every three Indian still lives in villages, the phenomenon of concentration of urban populations in large

---

[*] Akhilesh Surjan is Lecturer in Global Change and Sustainability Program, United Nations University – Institute for Sustainability and Peace (UNU-ISP), 53-70, Jingumae 5-chome, Shibuya-ku, Tokyo, Japan. Email: akhilesh.surjan@yahoo.com
[†] Prasanta Kumar Mohapatra is Project Engineer, Orissa Water Supply and Sewerage Board, Water Works Road, Puri (Orissa), India. E-mail: prasant_mohapatra@hotmail.com

cities and existing city agglomerations has led to tremendous pressure on civic infrastructure systems relating to water supply, sewerage and drainage, solid waste management and transport etc" (webindia123, 2009).

Urbanization in India is although remained slow and lower than the average for Asia, 286 million people living in India's cities today presents scores of challenges. The root causes of constantly growing urbanization have their origins in social, economic, political, institutional, environmental spheres. Complex amalgamation of opportunities and threats offered by cities has far reaching effects. It may be interesting to note the key observations (Box9a) from the 'India Urban Poverty Report 2009' which is a unique outcome of the work jointly done by the Indian government and United Nations Development Program. The 'urbanization of poverty' discussed in the report is altogether a fresh dimension breaking the myth which most modern cities carries with their 'utopian' impression.

URBAN DEVELOPMENT IN INDIA: EASIER SAID THAN DONE!

Sustainable urban development in India was not a matter of prime concern until recently as the country remained predominantly rural. Moreover, focus of the government for provision of housing and basic amenities for economically and socially backward sections of the urban society led to creation of private sector which solely catered to the needs of better-off population. Another segment which thrived in parallel is known as informal sector which found its niche between government and private and contribute most to jeopardize environment. Development planning at the national level is guided by Five Year Plans (FYP) which are drafted considering "the most effective and balanced utilization of the country's material, capital and human resources" (Surjan & Shaw, 2009). FYP of 1961-66 discussed for the first time, planned spatial development of major cities and promoted preparation of Master Plans to guide future development pathways of the major cities. Next FYP, which became functional from 1969, introduced radical changes where high priority was accorded to balanced urban growth. And as a result, Scheme for Environmental Improvement of Urban Slums was undertaken to provide a minimum level of services (i.e. water supply, sewerage, drainage, street pavements) in large cities. Congestion and overcrowding of large cities also were recognized during the year 1975. As a response to this, concepts like simultaneously developing small and medium cities gathered momentum after 1980.

Federal government established India's first National Commission on Urbanization (NCU) which issued its final report in August 1988 which brought a landmark change in the way cities were perceived until then. "The push factors like population growth and unemployment etc. and pull factors like opportunities in the urban areas are debated in the studies of India's urbanization and the NCU has termed them as factors of demographic and economic momentum respectively" (Bhagat, 2001). Starting 1992, India started opening its economy to the global markets and also FYP in 1992 explicitly recognized significance of urban sector in the national economy. It was also noticed that a large proportion of urban population lives in abysmally poor conditions and lacks basic services. Especially targeted employment programs for urban poor were introduced while also combining efforts to provide minimum access to services like water, sanitation, education, health and housing).

Box 9a. India releases its First Report on 'Urban Poverty'

India has issued its first-ever report on the nature and dynamics of urban poverty in the country. Undertaken with the support of the United Nations Development Programme (UNDP), *India: Urban Poverty Report 2009* identifies the problems faced by the poor and focuses on the systemic changes that are needed to address them.

The report examines various issues related to urban poverty, such as migration, labour, the role of gender, access to basic services and the appalling condition of India slums. It also looks at the dynamics of urban land and capital market, urban governance, and the marginalisation of the poor to the urban periphery.

Key messages of the report include:

- **Poverty in India has become urbanised.** Urban poverty in India is over 25 percent; some 81 million people live in urban areas on incomes that are below the poverty line. At the national level, rural poverty remains higher than urban poverty, but the gap is closing. By 2030, urbanisation in India is projected to reach 50 percent.

- **Migration towards urban centres has increased,** indicating that economic reforms have not been effective in creating jobs in small and medium towns as well as rural areas. Poverty was higher among rural to urban migrants, while the most successful migrants are those who move from one urban area to another.

- **Urban poverty poses different problems.** The nature of urban poverty poses distinct challenges for housing, water, sanitation, health, education, social security, livelihoods and the special needs of vulnerable groups such as women, children and the aging.

- **Slum populations are increasing.** According to a 2001 census, India cities have a slum population of 42.6 million (23.7 percent of the urban population). The majority (11.2 million) are in Maharashtra, whose capital city Mumbai is home to the Dharavi slum. While the slum population has increased, the number of slums has decreased · resulting in greater density.

- **Slum dwellers lack access to basic services.** Most slum dwellers do not have access to clean water, sanitation and health care facilities. They face a constant threat of eviction, removal, confiscation of goods and have virtually no social security cover. Some 54 percent of urban slums do not have toilets; public facilities are unusable due to a lack of maintenance. Some 54 percent of urban slums do not have toilets; public facilities are unusable due to a lack of maintenance.

Proposed solutions to urban poverty include a greater equity in the provision of basic services, targeted subsidies for vulnerable sections of the population, and special government assistance to strengthen the economic bases of small and medium towns. In slums, the report recommends organising slum communities, extending sewage systems and electricity to slum areas, and constructing public toilets that will be maintained by the community.

India: Urban Poverty Report 2009 is part of a UNDP-supported government project to develop a national strategy for urban poverty.

Source: Cities Alliance, 2009. http://www.citiesalliance.org/publications/homepage-features/feb-09/India_UrbanPoverty_Report2009.html

INDIAN CITIES AND THE ENVIRONMENT: A DELAYED CONCERN

Environmental degradation in Indian cities can be traced back for over a century or so however remained unnoticed until slums and squatters started taking a large chunk of cityscape and caused 'visual nuisance'. Until now, environment and forests are dealt in an integrated manner by the federal government leaving little room for considering deterioration of environment as independent entity. Single point agenda to control of air, water and noise pollution in the cities through legal provisions without creating conducive environment caused further dent in this short-sighted vision. Eco-City and Green-Community were the words considered alien to Indian context until the end of last century. Daily living practices of an ordinary citizen in India have always remained 'green', however, either not noticed or intentionally ignored. It may be noteworthy to look at the second annual survey conducted by the National Geographic Society and international polling firm GlobeScan on environmentally sustainable behavior (Box9b). The survey revealed that Indians as a group express above-average concern about the environment and practice environmental friendly living compared to most economies surveyed.

Box 9b.Indians are world's 'Greenest': Survey

Indians may be green with envy at the consumption-driven lifestyle in the West, but their own frugal ways and modest means have catapulted them to the top spot in the world's Green index, making them the most environmental-friendly denizens of Planet Earth.

The second annual survey conducted by the National Geographic Society and international polling firm GlobeScan on environmentally sustainable behaviour, the results of which were released on Wednesday, showed that Indian consumers have overtaken Brazilians to take the top spot with a Greendex score of 59.5. The Chinese retained the third spot with 55.2. At the bottom of the ladder in the 17-country survey are over-consumptive Americans (43.7), Canadians (43.5) and Japanese (49.3).

So, what has put Indians at the top of the Green ladder? It was driven by above-average performance on all four sub-indices, including first-place rankings for food and goods. Indians are the most frequent consumers of self-grown food, with 35% eating what they grow several times a week or daily. Desis are also the least frequent consumers of beef, which requires greater energy to grow—only 22% consume beef weekly compared to an average of 63% for the 17 countries surveyed.

Indian consumers also topped the goods sub-index score. Their top status is due in part to having lower-than-average rates of ownership of large appliances and electronics. Also, the rate of those buying used goods, avoiding environmentally unfriendly products and excessive packaging, and buying environmentally friendly products is the highest.

Indian consumers continue to rank third on the transportation sub-index, based on the fact that they are second-most likely to live close to their usual destinations and second-least likely to own a car or truck (54%). Among those who drive, Indians tend to have lower-than-average annual mileage rates. Besides, they are the most likely to own and use motorcycles or scooters and second most likely to drive a compact car, after the Mexicans. In addition, walking or riding a bike is up seven points from the past year (to 57%).

As regards housing, Indians rank second only to Brazilians. Factors contributing to their high ranking include a low incidence of having home heating (41%) and hot running water (38%) and a high incidence of using on-demand electric water heating (45% among those with hot running water), and using solar energy to heat water (15%).

However, there are plenty of signs that India's Green-ness, which seems driven more out of compulsion than conviction, may not last long. According to the study, Indians are the most likely to say that they intend to acquire a motorised vehicle in the next year (58%). There is also declining frequency of consumption of local foods, fruits and vegetables and an increase in the consumption of imported foods and bottled water. India is also the only country surveyed experiencing increased bottled water consumption.

Indian consumers' attitudes showed their divergent and conflicting views on the environment. As a group, they express above-average concern about the environment. As for their personal contribution to environmental problems, they say they are trying hard to reduce their own negative impact and are paying more attention to environmental issues. At the same time, many agree that environmental problems are exaggerated and that the Green movement is a fad. Indians have faith in the government, industry and new technology to help solve environmental problems, but express below-average faith in the ability of individuals to make a positive impact.

Source: Times of India, 14 May 2009. http://timesofindia.indiatimes.com/NEWS/ Environment/Global-Warming/Indians-are-worlds-greenest-Survey/articleshow/4527041.cms

It is interesting to look at the government policies and interventions towards urban development and environment as well as behavior of masses which provides important links to envision Eco-City in Indian context. Next part of this paper will give brief account of Eco-City development efforts in selected Indian cities and will later focus on the City of Puri for greater insights.

ECO-CITY IN INDIA: GLASS HALF EMPTY/HALF FULL!

In the beginning of year 2000, a German bilateral agency introduced the idea of 'Eco-City' to India's federal agency responsible for 'environment and forests' (Ministry of Environment and Forests (MoEF). It may be noted here that there is a dedicated federal agency in India, known as Ministry of Urban Development and Poverty Alleviation (which is later renamed as 'Ministry of Urban Development' - MoUD) which looks after urban issues and ideally the concept of 'Eco-City' should have been routed through it. Nonetheless, the Eco-City project was introduced by MoEF through its statutory organization known as Central Pollution Control Board (CPCB). It is interesting to look at origins of CPCB which was constituted in the year 1974 under the Water (Prevention and Control of Pollution) Act, 1974. Further, CPCB was entrusted with the powers and functions under the Air (Prevention and Control of Pollution) Act, 1981 (CPCB, 2009).

It may be argued that why a statutory body entrusted with looking after pollution was at the center-stage to coordinate 'Eco-City' ideas. Despite efforts, it was not possible to get clarity of this important institutional dimension. Nonetheless, a close look at some of the Annual Reports available from the official website of CPCB offers some description of beginning of the concept of 'Eco-City' in Indian context and its status in the span of about 6 years, as follows:

1. Eco-City Program Description at CPCB Website

"Pilot studies conducted for urban areas ....at the CPCB under the World Bank funded Environmental Management Capacity Building Project and supported by the GTZ-CPCB Project under the Indo-German Bilateral Programme, were encouraging. Using the experiences from the pilot studies conducted for urban areas, the Ecocity programme was conceptualized for improving environment and achieving sustainable development through a comprehensive urban improvement system employing practical, innovative and non-conventional solutions. ..... The overall objective of the Ecocity project (Scheme under the 10th FYP) .... is to improve environment and bring in visible results through implementation of identified environmental improvement projects in the selected towns/cities. The specific objectives are to: (a) Identify the environmental problems/hotspots in the identified towns and priority environmental improvement projects through participatory approach; (b) Designing & detailing the prioritized environmental improvement projects; and creation of landmarks that shows visible environmental improvement". (CPCB, 2009a)

2. Eco-City Definition, Brief Background of its Origin, as Defined in Annual Report 2001-02

"'Ecocity' is a city that decreases environmental burden/stress, improves living conditions and helps in achieving sustainable development through a comprehensive urban improvement system involving planning and management of land and its resources and implementation of environmental improvement measures. The Ecocities include area-wide improvements and providing of infrastructure and services. The Zoning Atlas team of CPCB provided the concepts and technical support for the pilot Ecocity project for the Kottayam-Kumarakom Region of Kerala. The project was launched in January 2002 by the Ministry of Environment & Forests, Government of India. A pre-feasibility study was undertaken with GTZ experts, under the Indo-German Bilateral Programme, on developing "Taj Ecocity" around Taj Mahal with verifiable environmental improvement considering social, economic and environmental aspects". (CPCB, 2009b)

3. Eco-City - Project Based Approach in Small/medium Towns, as Defined in Annual Report 2002-03

"Ecocities programme is being extended to small and medium towns in the country. The programme is aimed at bringing in verifiable environmental improvement through execution of identified projects. The work has commenced for launching the project in Mathura (Uttar Pradesh), Ujjain (Madhya Pradesh), Puri (Orissa), Vapi (Gujrat), Thanjavour (Tamil Nadu),

Rishikesh (Uttaranchal), Tirupati (Andhra Pradesh), Shillong (Meghalaya), Kottayam (Kerala) and Vrindavan (Uttar Pradesh)" (CPCB, 2009c).

4. Eco-City for Visible Environmental Improvement on Cost Sharing Basis, Annual Report 2006-07

"The Eco-city Programme has been conceptualized for improving environment and achieving sustainable development through a comprehensive urban improvement system employing practical, innovative and non-conventional solutions. Under the project, funds will be provided to the Municipalities by Central Pollution Control Board for the identified project, on 50:50 cost-sharing basis up to a maximum of 25 million Indian Rupees (approx USD 513,000 with prevailing conversion rate on 18 August 2009) per town, wherein 50% of the total budget should come from Municipalities either from their own funds or through financial institutions or any other means. The following towns have been taken under first phase of Eco-city programme to bring in visible environmental improvement: Vrindavan (Uttar Pradesh), Tirupati (Andhra Pradesh), Puri (Orissa), Ujjain (Madhya Pradesh), Kottayam (Kerala), Thanjavour (Tamil Nadu)". (CPCB, 2009d)

It is apparent from the above journey through time that involvement of pilot studies done through involvement bilateral experts and international assistance encouraged to further expand Eco-City concept in India. However, rather than transforming the city as a whole into Eco-City, identifying 'visible' environmental improvement project in a participatory manner and implementing them on a cost-sharing basis was the approach followed to realize the Eco-City dream.

There are number of inadequacies which have weakened the Indian version of Eco-City even before it can take shape. First of all, the philosophy behind Eco-City was not understood in local context. Indian 'Green-Practices' were neither studies nor highlighted. Using the word Eco-City itself is extravagance because the scale of consideration was rather limited from the beginning with 'identifying few projects in a particular city', instead of 'whole city as Eco-City'. There seems to be little seriousness involved while considering local governments as cost sharing partner because most local governments are suffering from serious resource crunch and have multitude of problems which paralyses day-to-day urban management tasks. In the following part of the paper, finer details of the Eco-City project are discussed in the context of a city of Puri. This will help understanding the whole discussion and argument in a localised context.

PURI: AN INTRODUCTION

1. Historical Geography

Puri is an ancient town in the state of Orissa (India), with a history dating back to the 3rd century B.C. It is famous for its historical antiquities and religious sanctuary. The importance of the town as a seat of *Vaishnavism* increased when the temple of *Purusottam Lord Jagannath, Lord Balabhadra* and *Devi Subhadra* was constructed. Although Puri was an ancient habitation since time immemorial, it was officially notified as a town only from 19th January 1865 by the British Government. Subsequently, Puri Town was declared as a Municipality with effect from 1st April 1881. The Orissa Municipal Act 1950 was extended to Puri with effect from the year 1951. Presently, Puri Municipality consists of 30 wards.

Mythologically, the city believed to have been planned in the shape of conch or 'shankha-ksehtra'. This area is about five *kroshas* (10 miles) in extent of which nearly two *kroshas* are submerged in the sea. In the center of the *kshetra* lies the hillocks called *Nilagiri* where the temple stands and within its compound stand many sacred institutions. The broad end of the *Shankakshetra* lies to the west where the temple of *Loknath* is located and on the apical end or tail end in the east is located the temple of *Nilakantha*. The town consists of the Temple of Lord *Jagannath* in its centre with eight quarters (Sahis) radiating from it. The town colonies were called *'Sahis'*; each *Sahi* had people of particular caste. The population of the town composed mostly of the officiating priests and officers attached to the various activities of the temple and were dependent on the temple. Besides, there were inmates of the *mathas* which chiefly clusters round the temple. And as the rituals of the temple and *Ratha Yatra* (Car Festival) become more elaborate, more people started staying here. By this time ordinary people who had no contribution to rituals had also started staying. The town was originally built on sand and probably began from north and west towards Cuttack and Ganjam road where the *Madhupur* or *Matia* stream is lined with the largest tanks, *Narendra, Mitiani, Markandeya* and *Siva Ganga*. As the tributary of the river *Matiani* draining into the sea dried up, it paved the way for the new direction of development of the town in the southeast.

The *Matha* Complexes with their system of interconnected open courtyards as well as extensive areas of garden and plantations were also closely associated with the major temple complexes. These *Mathas* were located within compact traditional neighborhoods or *'Sahis'* with a distinctive built form, which emerged in close proximity to the *Jagannath* temple complex. The evolution of the town is believed to have begun with *Harachandi Sahi & Baseli Sahi* in the northwest, *Kundhei-Benta Sahi & HeraGouri* on the north-east, *Markandeswar Sahi & Chudanga Sahi* on the north; then west towards *Lokanatha, and* later to the east and south with *Bali Sahi* on the south, with *Matimandapa Sahi* and *Kalikadevi Sahi* on the south-east towards *Swargadwar* and Goudabada Sahi on the south-west, and *Dolamandapa Sahi* and *Mochi Sahi* on the east. This was followed by the *Badadanda* area towards the *Gundicha* Temple. However, large areas of garden and plantations were integrated with the traditional *'Sahis'* and large open spaces were also associated as a green buffer around to sustain them. The sandy tract known as *Balukhand* was not constructed upon, with the exception of the *Swargdwar* area where certain *Mathas* and a cremation ground were established, maintaining a distance from the seashore.

2. Evolution of Jagannath Temple Complex

Since the last one and a half centuries, different historians have expressed different opinions about the time when the present temple was constructed. Stenling in his 'History of Orissa' paper writes that *Ananga Bhima Dev* built the temple in 1196 A.D. Fergusson in his 'History of Architecture' (Volume II, Page 592) opinioned that the temple was built in 1098. William Hunten in his 'History of Orissa' writes that the construction of the temple began in 1114 and finished in 1198. This fits in with the popular tradition in Orissa that it was *Ananga*

Bhima Dev, who constructed the present Temple and fixed the various *sebas, sebaks* and *Nitis*. According to the palm leaf chronicles (*Madala Panji*) of Puri temple, *Yayati Keshari* erected a temple for Lord Jagannath but no remains of this temple is in existence. It is said the present temple of lord *Jagannath* is of the 12th Century King *Chodaganga Dev* (1112-1148 A.D) of the *Ganga* dynasty who began the construction of the present temple sometime after 1135 A.D. The temple was completed by his worthy descendant *Ananga Bhima Dev*. Thus the city of Puri became the religious metropolis of Orissa and of India.

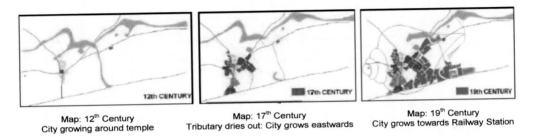

| Map: 12th Century | Map: 17th Century | Map: 19th Century |
| City growing around temple | Tributary dries out: City grows eastwards | City grows towards Railway Station |

Figure 9a. City of Puri.

3. Present City of Puri

Puri is located between 190 47' and 190 50' North latitudes and 850 48' and 850 52' East longitudes on the shore of Bay of Bengal. It is 65 Km away from Bhubaneswar, the administrative capital of Orissa State. The town is connected by broad gauge railway line to most parts of the country. Present day Municipality limits spreads over 16.26 Square kilometer land. Puri is considered a holy place. The present (2008) population is estimated to be 186,000. The town's resident population is growing exponentially (Figure 9b) since 1941 and has not stabilized yet though land available is restricted. Devotees and tourists come to Puri all round the year. They offer prayer, enjoy the pristine beauty of sea and relax. Over the years, floating population is increasing. Average number of tourists visiting the town per day is 15,000. On festival days and New Year day, the floating population increases to 500,000. Estimates suggest people visiting Puri on Car festival day, a festival celebrated each year on a pre specified day, reaches a peak of 1.0 million. Visitors during car festival recorded in 1841, 1892 and 1901 were 112,000; 200,000 and over 300,000 respectively. In year 2008, on Car festival day held on 4th July 2008, about 800,000 people visited Puri. It is projected that in year 2015 on this festival day, the town must be ready to accommodate 2.5 million guests. The coast line of the town is about 6.59 Km long. Many leading hotels have been constructed in recent years by private entrepreneurs along the beach boosting tourism (Figure 9c).

The climate of Puri in summer is warm and humid with average maximum and minimum temperatures of 37.5^0C and 27^0C respectively. The climate in winter is cool and pleasant with average maximum and minimum temperatures of 28.2^{0C} and 15.2^0C respectively. About 74% of the rainfall is received during monsoon season from June to September. Maximum precipitation occurs in July. On an average, there are 71 rainy days in a year. Annual precipitation in Puri is about 1520 mm. Puri experiences cyclonic depressions and storms originating in the Bay of Bengal.

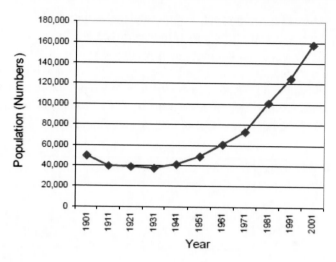

Figure 9b. Population of Puri.

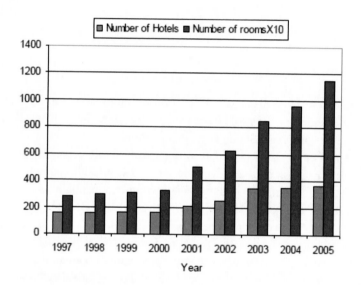

Figure 9c. Hotels have been constructed in recent years.

The population living in peri-urban area often called slum constitute about 25% of town population. A slum is characterized by poor housing quality with lack of access to basic amenities. About 33,768 persons live in 6753 houses in 46 slum pockets (2001 census). A majority and highly populated slums exist near the sea shore. Traditional occupation of these slum residents are fishing in sea.

From the foregoing description, it can be summarized that Puri is a vibrant town with temple and sea beach being main drivers of population and economic growth. The town is witnessing phenomenal increase of resident and tourist population. The pleasant weather, traditional rituals and numerous festivals contribute to seasonal peak in tourism. Tourism and related activities provide employment to people. The town's slum population is a cause of major concern as far as development is concerned. As resident and tourist population are increasing at fast rate, there is bound to be pressure on the existing infrastructure to provide urban basic services commensurate with town's status as a major tourist hub.

DISCUSSION ON KEY ENVIRONMENTAL STRESSORS OF PURI

Puri today experiences a number of environmental stresses and they elements not separable rather interdependent in a complex amalgam. However, in the following part, only three key stressors are discussed in detail. These indicators are selected on the basis of general community perception as well as strong opinion of local government. The three prominent indicators of environmental stress in Puri are water supply, sanitation and solid waste management.

1. Water Supply

Puri town is bestowed with nature's gift in the form of two groundwater aquifers located on the eastern and the western side of the town. The domestic water supply since the year 1935 is met from groundwater. The waterworks areas are locally known as the *Baliapanda* and *Chakratirtha* waterworks. To meet domestic water need, wells have been installed in these two water fields. Some wells have been installed within the town. Currently, about 26 Million liters groundwater is drawn daily through production wells and hand pumps etc. The groundwater is of very good quality requiring no treatment. Post-chlorination is practiced to maintain residual chlorine in the distribution system (CDP, 2006).

The pressure of urbanization on the land reserved for waterworks area as drinking water source is visible. A slum settlement has proliferated close to the groundwater field at *Baliapanda* water field, which lies on the western boundary of the city. An inter-state bus parking has come up at another *Chakratirtha* water field at the eastern boundary of the city. While trying to solve the transport related problem, a water quality related problem has been created inadvertently. The two aquifers are already under stress due to anthropogenic activities. The consequence of these activities is pollution of water sources. The recent study by National Environmental Engineering Research Institute - NEERI (NEERI, 2008), shows that water security is under the threat of anthropogenic activities in the town.

In order to prevent pollution of the water sources, the practical approach followed by "Berlin Water Supply" is worth emulating. Berlin City meets 78% of its domestic water requirement from groundwater. The waterworks' production wells are protected from pollution and for this purpose, the area surrounding a well is divided into three zones and certain activities and uses are forbidden in these zones (Box9c). Puri town can adopt the Berlin Water's practice following a scientific approach to the problem. The town is situated near the coast in which the fresh water and salt water are in delicate balance. The fresh water availability in Puri town is complicated by over pumping near the coast resulting in sea water intrusion. Also, the emergent climate change scenario can no longer be ignored.

Box 9c. Groundwater protective Zones and differentiated functions

The Berlin waterworks area, known as water conservation district, has protective zones with differentiated use restrictions. For example, Zone-I extend up to a radius of 10 m from a well. Zone-II reaches from the border of Zone-I to a line at which groundwater requires 50 days to flow into the well. Zone-III comprises the area from the outer border of Zone-II and the border of the groundwater catchment area. For large catchment exceeding 2 Km, Zone-III is subdivided into Zone-III(a) extending up to 500 days and Zone-III(b), up to 2500 days (Berlin Water, 1995).

Zone-I prohibits:

- All activities except for those measures required for water supply purposes.

Zone-II prohibits:

- The erection or alteration of constructed facilities (residential buildings, industrial facilities, streets) with the exception of alterations in buildings.
- The erection and operation of facilities for dealing with substances harmful to water and the transportation of substances harmful to water through pipelines.
- The discharge of wastewater and non-purified rain water into surface waters.
- The use of fertilizers, insecticides, herbicides and pesticides.
- The removal of water and solid materials from the subsoil and from ground surface.

Zone-III prohibits:

- The discharge of wastewater into the subsoil with the exception of the spraying or seepage of rain water through the topsoil.
- Activities which enable the penetration of pollutants into surface waters, the subsoil or groundwater; in particular the dismantling and repair of motor vehicles, including the changing of oil on un-stabilized soil.
- The erection, rebuilding, extension or basic alteration of buildings, unless there is a complete and proper wastewater drainage by sealed pipelines into public waste water facilities, or the collection of drainage in permanently sealed pits with proper disposal.
- The erection and operation of dumps for waste harmful to water and/or waste disposal facilities which require planning approval.
- The construction of gravel pits, the placement and storage of materials harmful to water directly on the subsoil, particularly for the construction of streets, canals, and paths, with the exception of small improvements with no danger of contamination.
- The construction or extension of drains for dewatering the subsoil.

Berlin has a total surface area of about 890 sq. km. and 28% area accounts for water conservation zones.

2. Sanitation

In the town, houses have water flush toilets connected to septic tank and soak pit system. Groundwater table is high in major part of the town which almost reaches ground surface during the rainy season. Some parts of the town remain submerged in water following a rain. The older part of the town has drains on side of the street pavement to discharge rain water. The drainage network can be described as a combined system and carries both rain water and wastewater. It is not unusual to find organic matter floating in drain water. The drain transports large amounts of solid wastes which are dumped in to it.

Infiltration of waste water in the sandy strata on which the town has grown leads to groundwater pollution. The major sanitation issues can be outlined as follows:

- Rain water drains function as combined system conveying rain water and wastewater.
- The problem of silting of drains, filling with garbage and solid waste.
- The town lacks a drainage system and a sewerage system. High water table during rainy season causes unsanitary condition.
- Septic tank and soak pit causes infiltration in sandy formation leading to pollution of groundwater.
- Poor hygiene practice by slum population, open defecation and defecation alongside the drains.
- Inadequate public amenities for the floating population.
- Water bodies inside the town are very old and have religious significance need renovation to check water quality deterioration and improve water quality.

Of all the issues highlighted above, pollution of groundwater and the sea beach are chief concern. Pollution of water source is undesirable for protection of public health. Pollution of beach and the sea water will dissuade tourists from visiting the town. The economy of the town will be hurt.

3. Solid Waste Management

The generation of solid waste in a town is influenced by the socio-economic status and attitudes of its people. Puri produces 79 Ton of solid waste daily. The major contributors are the temple, hotels and restaurants on the sea beach and the market places. The older city having narrow lanes poses challenges for effective collection of waste. About 50 Ton is transported daily to a solid waste treatment plant located inside the town. The plant is close to "*Baliapanda* water work" on the western edge of the town (Bisoyi, 2004).

The important solid waste management issues are:

- Separation of waste at source.
- Innovative collection mechanism considering that most streets are narrow and congested.
- Keeping the temple surrounding, market area and sea beach area clean.
- Recycling of waste.
- Public participation and awareness.

For Puri, these issues are important to keep the town clean and attractive for tourists. The existing collection system is inadequate and requires strengthening. Solid waste often finds its way to public sewer and obstructs flow in drain. One of the methods adopted by the Puri municipality to dispose of the household waste and other solid wastes is to landfill in vacant spaces inside the city little realizing that this has adverse impact not only the neighborhood but also on the groundwater quality. The experiences of the project partners from the two demonstration pilot projects implemented to strengthen the solid waste management system in Puri under Ecocity as well as European Union cooperation project are given in Box 9d.

Box 9d. SWM demonstration project by GTZ and EU

AAWam by ASEM, GTZ (GTZ, ASEM, 2008):

The Project on "Achieving Action in Waste Management" (AAWaM) was formulated by the Ministry of Environment & Forests (MoEF) together with GTZ under the ASEM Programme and with the Centre for Environment Education (CEE) under the United Nations Development Program – Small Grants Program (UNDP-SGP-GEF). The project has the objective of identifying and implementing environmentally and economically sustainable strategies for municipal solid waste management in the identified cities of the EcoCity Project. It aims to generate awareness and facilitate community participation in solid waste management in the cities including involving NGOs to work in the identified demonstration areas and in establishing appropriate technologies to maximize the recycling and reuse for the selected streams of waste. Four towns namely, Tirupati, Puri, Ujjain and Vrindavan are covered under the project.

In Puri city, with the efforts of CEE team, Resident Welfare Associations (RWA) have been formed along the street adjoining the Jagannath temple and NGOs have been identified for door -to-door collection of waste segregation and removal from the area. In March 2008, the area served in Puri by the NGO - M/s F.M.Welfare Club was around Shree Jaganath Temple - from Singadwar to Market square to Narendrakona; Chakratirtha road and Atharanala area and the Sea Beach area (i.e. from Digabarini square to hotel Hans coco Palm, which is approximately 3km more area). This additional area has been allotted by the Puri Municipality to M/s F.M.Welfare Club without the process of tendering. Of the total 1138 beneficiaries under the project, there were 311 households, 442 commercial establishments/ shops, 26 educational institutes/ offices, 215 hotels/lodges and 144 informal sector vendors. Employment has been provided to 29 workers of whom 20 are rag pickers and 2 are women. 22 Resident Welfare Associations and 7 partnerships have been formed at various levels in the city.

Solid Waste Management in Puri:- A EURO-ASIA cooperation (INARE, 2008)

In this EU funded project, a partnership mechanism was developed in Puri for a pilot project of solid waste management. The project included exchange of experiences, best practices, expertise and information between European Union and Indian municipalities, the development of community involvement in waste management, awareness-raising events, educational materials complemented by seminar and training events, enabling technologies and activities within the Indian cultural context to improve the environmental quality. The project effects are the following:

- The Solid Waste Management programme trained 3000 persons from different target groups at Puri.
- Additional equipments were provided like waste bins of different size, trucks for collection, clothing and face masks protection for the cleaners
- Waste Management Groups from beneficiary community were established to manage the programme sustainably.
- Waste segregation at the source by the community was improved by their involvement in devising the scheme through improved public participation.

> • Contribution and co-operation in the management of the system were improved.
>   The project aimed to improve and strengthen waste management systems within Puri district through public participation and improved mechanism, co-ordination of industry and civil societies to strengthen Europe-Asia cooperation. Initially, key groups within the population were targeted. These included municipality personnel involved with community consultation and developing waste collection systems, key business sectors namely hotel personnel, local farmers, students and young people, women and local non-governmental organizations.

The SWM demonstration pilot projects involving mostly NGOs had marginal effect. The experiences could not be replicated in other areas of the town and the NGOs could not be substitute for the mammoth task carried out daily by the municipality. The interest of the local NGOs dwindled once the project proponents were out of the scene.

It is seen that in the current situation, issues relating to sanitation and solid waste management in Puri town described in preceding sections adversely impact groundwater quality. Other potential impact on groundwater quality is sea water intrusion, which may occur if quantity of groundwater withdrawn is more than safe quantity. Over pumping may occur if exact potential of aquifer is not known. Pollution of beach is another negative effect if untreated or partially treated wastewater is discharged to sea.

PURI AS ECO-CITY: AN ABORTED EXPERIENCE

CPCB in association with technical experts and organizations did assessment studies and prepared Eco-City Plan of Puri. The Plan discusses wealth of issues and recommends a big list of projects to be implemented for bringing about change. However, due to financial constraints have forced to select only top 5-6 projects to make a visible impression. Moreover, institutional ambiguity over the fate of the Eco-City project remained large as in the end of year 2005, the federal government launched a bigger, more comprehensive project known as '*Jawaharlal Nehru* National Urban Renewal Mission (JNNURM)', with big chunk of money. Urban Development ministry of federal government introduced this project which has much larger and direct say in the matters of urban development and planning for the future of the cities in India. Assistance of US $12.5 billion for integrated planned development of 63 selected cities in India, including Puri was a big boost which is now receiving for improvements in infrastructure development, urban poverty and urban governance in these cities (Surjan and Shaw, 2008). It is much more affordable to the local government which is required to share only 10% of the total bill, the municipality of Puri moved ahead to receive JNNURM support and is expected to implement several projects identified through a much detailed study known as the 'Puri City Development Plan' (PCDP).

In other words, Eco-City project in Puri was almost superseded by much larger project. The priority projects identified by the Eco-City Plan were: (a) rejuvenation of Markendeya Tank (b) Repair/Improvement storm water drainage around Jagannath Temple (c) Repair/renovation of public toilets (d) Repair of renovation of drinking water facility near Jagannath Temple (e) Development of Eco Automobile Park (f) Demonstration pilot project to improve municipality solid waste (MSW) management.

Though the Eco-City project began in 2002, excepting the demonstration project for MSW management, other work components were not grounded in Puri city yet. The issue here is not to analyze the reasons for non-implementation of these works, but in the context of the aims advocated in the Eco-City project, i.e., to improve the existing environment, addressing the factors responsible for environmental damage and protection of environmental resources, it is worthwhile to assess whether these works would make Puri truly an Eco-City. In so far as water and sanitation issues of the town are concerned, the Eco-City project has chosen to implement renovation of few water faucets near the temple, improvement of two existing toilets near the temple and reconstruction of storm water drains around the temple.

It can be noticed that Eco-City project components for Puri are not truly holistic as far as improvement of environment is concerned. For example, the renovation of a drain surrounding the temple is not the panacea for achieving sustainable sanitation system in the town i.e. provision of drainage and sewerage system for improvement of environment and health of the people. The improvement of a drinking water facility/faucet near the temple does not solve the problem at source. The MSW pilot project failed to notice unsanitary solid waste landfill sites existing within the city. In fact, one of the pilot projects created a new dump site in a water field area.

Epilogue

This paper attempted to identify some environmental issues of concern in Puri, a coastal town. The issues relating to water use, sanitation and solid waste management were highlighted as they are interlinked due to the geography and location of the town on the coast of the Bay of Bengal. Some institutional perspectives about Eco-City development approach in India are also highlighted. Eco-City models in India is yet to be realized holistically and focus merely on selective infrastructure development or upgrading should not be ill-represented as 'Eco-City'. Present half-hearted attitude towards Eco-City in Puri not only hampers the innovativeness and ideological constructs but also ruins the 'Eco-City' vision among people and policy makers alike. In order to make Puri a true Eco-City, it is necessary that a comprehensive social, political and technical mechanism be adapted to understand and solve various challenges while hand-in-hand improving the urban environment.

With the looming climate change impacts which are likely to worsen the present environmental stresses, bring in more environmental refugees by migration from rural to urban areas, increase likeliness of catastrophic extreme events and seal level rise, it is still not too late for Indian cities to reinvent Eco-City paradigms. Local identification of existing community and institutional environmentally responsive habits needs preservation while gradually eliminating wasteful ones. This will help building resilience to environmental stresses and create buffer against global climate changes, thus, demystify illusion of Eco-City in India and elsewhere in the world.

ACKNOWLEDGEMENTS

The authors gratefully acknowledge the encouragement of the Chief Editor to contribute this paper. The paper would not have been possible without the cooperation of local government officials and also scientists from the Orissa State Pollution Control Board. The authors would like to convey their sincere thanks to all those who contributed to raise their understanding of this subject by sharing bits and pieces of information available in disperse form in print, web and handwriting.

REFERENCES

Berlin Water. (1995). *"Berlin Water Conservation Districts and Groundwater Use (Edition 1995), Section 02.11"*. Downloadable from:
http://www.stadtentwicklung.berlin.de/umwelt/umweltatlas/eda211_03.htm

Bhagat, R.B. (2003). *Urbanisation in India: A demographic reappraisal.* XXIV IUSSP (International Union for the Scientific Study of Population) General Conference, August 2001, Brazil. Available at http://www.iussp.org/Brazil2001/s80/S83_03_Bhagat.pdf accessed on 30th June 2009.

Bisoyi, L.K. (2004). *"Solid Waste Management in Puri Municipality- Problems and Prospects- a case study"*. Paper published in the 45th Annual Technical Session Volume, Institution of Engineers (India), Orissa State Center, 15th Feb. 2004.

CDP (2006). *Puri City Development Plan, 2006.*

Cities Alliance (2009). *Indian releases its first report on urban poverty.* Available at http://www.citiesalliance.org/publications/homepage-features/feb-09/India_Urban Poverty_Report2009.html accessed on 27 June 2009.

CPCB, (2009). Introduction. Official Website of the Central Pollution Control Board available at http://www.cpcb.nic.in/Introduction.php accessed on 18 August, 2009.

    a) CPCB, 2009a. Eco-City program. Official Website of the Central Pollution Control Board available at http://www.cpcb.nic.in/Urban_Environmen.php accessed on 18 August 2009.

    b) CPCB, 2009b. Annual Report 2001-02, page 143. available at http://www.cpcb.nic.in/upload/AnnualReports/AnnualReport_6_annualreport2001-02.pdf accessed on 18 August 2009.

    c) CPCB, 2009c. Annual Report 2002-03, page 135. available at http://www.cpcb.nic.in/upload/AnnualReports/AnnualReport_7_annualreport2002-03.pdf accessed on 18 August 2009.

    d) CPCB, 2009d. Annual Report 2006-07, page 225. available at http://www.cpcb.nic.in/upload/AnnualReports/AnnualReport_34_final-report-06-07-A.pdf accessed on 18 August 2009.

GTZ-ASEM. (2008). EcoCity Project - Pilot Efforts on Achieving Action in Waste Management, GTZ-ASEM, July 2008.

INARE. (2008). Technical final report of the project "Solid Waste Management in Puri: a Euro-Asia Cooperation" at final Conference "A clean city for healthy citizens", INARE, March, 2008.

Indicator Puri, profile of Puri http://www.ecocities-india.org/Downloads/index_eng.html

Mohapatra, P.K., (2008). *Environmental Impact of Water and Sanitation Issues.* Paper presented in the Coastal Cities conference in USA.

NEERI. (2001). "Sewage Collection and Treatment Systems for Puri", Final Report. NEERI, Nagpur, October 2001.

NEERI (2008). Study of groundwater aquifer system at Jagannath Puri and its protection from contamination, NEERI, Nagpur.

Surjan Akhilesh, Shaw Rajib. (2009). 'Institutional perspective on interlinking Environment and Disaster Management towards Sustainable Urban Development: Case of Orissa (India)'. Asian Journal of Environment and Disaster Management, (June 2009). http://www.rpsonline.com.sg/journals/
ajedm_toc.html

Surjan Akhilesh, Shaw Rajib. (2008). 'Eco-city' to 'disaster resilient eco community': A concerted approach in the coastal city of Puri, India'. Sustainability science, (October 2008). Springer publication. http://www.springerlink.com/content/u6n1ll183246048n/.

Times of India, 14 May 2009. *Indians are world's 'greenest': survey.* Available at http://timesofindia.indiatimes.com/NEWS/Environment/Global-Warming/Indians-are-worlds-greenest-Survey/articleshow/4527041.cms accessed on 14th May 2009.

Webindia123, 2009. *Approx 10% rise in proportion of urban population in India.* New Delhi, Jul 9 2009 available at http://news.webindia123.com/ news/Articles/India/20090709/1291 589.html accessed on 10th August 2009.

In: Eco-City and Green Community
Editor: Zhenghong Tang

ISBN: 978-1-60876-811-0
© 2010 Nova Science Publishers, Inc.

Chapter 10

INTERNATIONAL CASE STUDIES OF GREEN CITY AND URBAN SUSTAINABILITY

Ting Wei and Zhenghong Tang
Community and Regional Planning Program
University of Nebraska, Lincoln, NE, USA

Many old cities have manifested a culture of sustainability from generation to generation. Historically, many traditional cities grew and prospered by having a sustainable supply of food and other products from the surrounding countryside. Some old China cities utilized waste matter as fertilizer for agricultural production which helped to maintain a self sustainable system. Medieval European cities such as Siena or Dinkelsbuehl, as well as many Asian cities, had concentric rings of market gardens, forests, orchards, farm and grazing land. In some areas this situation continues today. Mont-Saint-Michel is a typical early example of the "Dense City" model. Located in northern France, it has stable high density from center to boundary and a distribution of services and facilities throughout the whole city. It was a "walking city" with short distances making for intensive social interaction. That is, work places, living areas and public places or social places are near each other or overlapping. Such cities have high population densities which encourage a social mix and interaction, the major characteristic of traditional cities.

The story of Tokyo is equally as interesting so it will be useful to discuss it in some detail. It was originally called Edo; its history dates back to the 10th century BC. Novelist Eisuke Ishikawa has researched the Edo Period and describes it as a highly sustainable society. He found that when Edo was first established, great piles of rubbish disfigured it and created very unhealthy conditions. However by about 1790, the residents of Edo had created an extremely clean, hygienic and sanitary urban environment. Nevertheless, in the absence of mass production and consumption, the economy grew by only 0.3 per cent a year. The sustainable utilization of limited resources in a continuous circular system had become the norm with widespread reuse and recycling of waste materials. "End-of-life" goods were not discarded; instead, tens of thousands of specialized traders and craftsmen were engaged in their reuse and recycling. They repaired old pans, pots and kettles. Ceramic repairers glued broken pieces of pottery. Others fixed tubs, barrels, lanterns, locks, inkpads, wooden footwear, umbrellas or mirrors. There were some 4,000 old clothes dealers in the city. Candle wax buyers even collected candle drippings and made them into new candles. This exemplified another type of sustainable city.

Modern cities can learn a great from these traditional cities. For instance, the compact city form can learn from the Medieval European city. Edo city is somewhat similar to the linear metabolism city model. However, present day cities cannot simply import unchanged traditional experiences into the 21st century. Urban planners must select their most applicable forms to incorporate into modern theory to build more sustainable cities.

Presently, most cities in the world have realized the importance of sustainable development. Some have made significant progress in developing sustainability in their environmental plans, transportation, residential areas, health facilities and sustainable initiatives. Such examples include Curitiba (Brazil), Portland (USA), London, UK, and Wellington (New Zealand). This essay chose Curitiba, Brazil and London, UK, for the following case studies since both have used modern techniques in their own way to achieve some degree of sustainability. Their eco- systems will be evaluated in terms of the above-mentioned qualities.

CURITIBA

Curitiba, the capital of Parana state, a Brazilian city of 1.8 million residents, is recognized as one of the most sustainable cities in the world; it has managed to incorporate a considerable variety of sustainable programs. Once it suffered from the common problems of high-speed expansion and shanty towns, but has now appeared as a leader among sustainable cities. It has become a global model of sustainability in the ecological, economic and social arena, even though part of a developing country. It has made sustainability and citizen participation the guiding principles of daily life, and the environment as its top priority. During his term as Curitiba's mayor, architect Jaime Lerner tackled its problems with a series of broad policies. As the shanty towns were mostly located on the banks of the city's rivers and lacked formal roads, nobody collected garbage which become massive fetid piles on the river banks. As a result, the river banks had no vegetation and the water was contaminated by sewage because of a lack of drainage. Lerner introduced a range of plans involving the participation of the shanty-dwellers to solve these problems. He offered transportation tokens to adults, and books and food to children in exchange for bags of rubbish delivered to the local dumps. Therefore, garbage was cleaned up and the area was landscaped. Rubbish collection, disposal and recycling became a core task for all citizens, reinforcing the ecological message the city wanted to promote. Curitiba cleaned up its rivers and then provided sewage treatment and turned the flood lands into large attractive parks. Twenty years ago, Curitiba had $1/2\,m^2$ per citizen. Today, after a systematic program of landscaping, it has a hundred times more, as well as a network of pedestrian and cycle routes.

Curitiba is particularly well known for its highly integrated bus system that services the entire city. Mayor Lerner said in a TV program (Rabinovitch & Leitman, 1996), *"if you want to help the environment try to do just two things. First: use your car less. And second: separate your garbage."* From the 1960s until the present, Curitiba created a network of bus services - some serving only local neighborhoods and other fast buses that run across the city on dedicated routes. By replacing conventional bus stops with so-called loading tubes, bus travel was greatly speeded up: as people enter the tubes they pay the resident conductor and when the bus arrives, everyone can get on and off instantly.

Other examples of Curitiba's and Lerner's vision include the transformation of the city's obsolete quarry into a landscaped cultural center plus three economical but inspiring, cultural centers. One building contains the "University of the Environment," built within a circular structure; here schoolchildren and their teachers follow a course which explains the principles and tangible results of urban sustainability. Lerner also commissioned a glazed opera house suspended over a lake, with a dramatic backdrop. Another of Mayor Lerner's accomplishments is a landscaped 25,000-person natural auditorium for concerts and festivals. Curitiba is robust rather than beautiful, but Lerner's urban agenda has created a genuine spirit of participation among its citizen.

Jaime Lerner (2000) said that in Curitiba there are many problems, as in other cities in Brazil and the world, but the main difference is the respect given to people. "We try to foster citizenship or co-responsibility – where the city and its residents share a common, sustainable target. The people understand that they can change the situation for the better if they act locally." Curitiba has developed strong social benefit programs to integrate its exploding population: nurseries and child care, schools, training, investment incentives, environmental protection and recycling. The extent and qualities of the eco-city system will be explained in the following text. This explanation will help readers quantify how and where the city achieved sustainability.

In the '60s, Agache, a French town-planner, attempted to impose a 1942 plan to widen the road system, demolish the borders of avenues and radially transform the city on behalf of the private car - just like North-American towns.

This plan was rejected by architects, engineers and by the Development Bank, which requested a study for a new and more realistic proposal. As early as 1966, a new plan was prepared by the Lerner team and was accepted, then frozen in 1971 under a harsh dictatorship. This latter plan closed the main road to private traffic; this annoyed some private interests. The earlier plan was concentric: to go from one district to another, traffic, both public and private, had to pass through the city center, which would certainly soon become congested. Therefore the streets had to be widened and the spiral of demolition and bottlenecks began.

The new plan was linear: the town was authorized to spread only along specified lines. The historical center, situated somewhat apart, could then become quietly pedestrianized. A ring-road connected the fast north-south and east-west bus routes and four concentric lines were added with stations that intersected the earlier lines.

The express radial routes would have needed a width of 60m which was impossible. The device adopted in the plan was to divide this flow intelligently between three neighboring parallel streets, the first and the third being one-way for private travel and the center being reserved for express buses and later for the tram or surface railway when the means became available. All this was coordinated with very little expropriation. The routes gave a structure to development without allowing it to occur anywhere at random and without creating impossible traffic conditions.

The first act of the local administration had been to look after the parks and to plant many trees. The inhabitants were mobilized with a slogan: "We bring shade, you bring fresh water" (an old Portuguese proverb). Previously the town had planted 5000 trees per year; this was now increased to 60 000 trees per year. In 20 years, Curitiba has increased the green space per inhabitant from 0.5[m.sup.2] to 52 [m.sup.2]. The intention was to plant one and a half million trees in 20 years for, since the 1988 murder of Chico Mendes (the campaigner to save the rainforest), Brazil has been on the defensive in ecological circles.

During a winter night in 1972 the work of pedestrianizing the main street began in greatest secrecy and was completed in 72 hours. In spite of the desire for participation, the preliminary work was carried out without publicity. The gamble succeeded. Participation was not by inhabitants but rather by urban guerrillas. The car addicts had decided to re-conquer the territory by brute force, but on Monday morning when the heavy equipment arrived to demolish everything, they were faced with a group of children painting on the ground. This was the first successful municipal sit-in for protecting pedestrians. Even so, it took two more years to put the express buses in operation, an essential accommodation for the pedestrians.

The Mayor organized popular well-constructed parks: the Iron Wire Opera, a round structure of completely glazed steel tubes, and a new Botanical Garden in which the greenhouses were also of steel tubes and domes. They were very popular.

Nothing would distinguish Curitiba from another town if it were not for the action of its mayor, architect Jaime Lerner. He has made all the difference. He quickly understood that a town is designed not by architects but by politics, so he contrived to be elected mayor for three alternate terms in 20 years and is ceaselessly improving the urban ecology of Curitiba.

In conclusion, Curitiba has incorporated most of the dense city model's principles: *"it should be a many-centered city, an overlapping activities city, an ecological city, an easy contacting city, an equitable city, an open city, and last, but not least, a beautiful city where art, architecture and landscaping can touch and satisfy the human spirit."* Meanwhile, it has proved that sustainability of a city can be measured by the eco-city qualities mentioned above.

LONDON

For four centuries London has been one of the world's most powerful financial, commercial and cultural centers. The legacy of this power and prosperity can be seen throughout the city: in its architecture, parks, squares, museums and public institutions. Even today London is rivaled only by New York in terms of extent and diversity of economic and cultural activities.

London was the first city to create a municipal administration capable of coordinating the complex matrix of London urban services which involved not only the public transport system, housing, water system and education, but also parks and museums. London's red double-decker buses, its police force and its advanced subway network, its schools and council housing, show a city committed to creating a humanist environment. This has been a great achievement. It is well known that only fifty years earlier London had the worst slums in the industrialized world - overcrowded, polluted and ridden with disease, where life expectancy was barely 25 years.

Subsequently, in 2000, an ecological footprint study for London called "City Limits" was recommended It calculated the energy used in agricultural production, transportation and processing, the land surface required for producing pet food and the sea surfaces required for fisheries. If these additional factors are included, the London footprint is actually 6.63 hectares per Londoner (Girardet, 2004). Compared to the UK average -- 4.4ha/person -- Londoners lifestyles can still be said to be unsustainable.

Therefore, Richard Rogers evaluated London's urban form using a dense city model and concluded that London is one of the least sustainable cities in Europe (Rogers, 1997). The book, *Towards an Urban Renaissance* (1999) outlined a sustainable city structure − the compact well-connected city form − which is the most important feature of the dense city. That is to say, the dense city has already incorporated the following strategies.

Nowadays, the stated vision is to "develop London as an exemplary, sustainable world city." A number of specific sustainability plans and actions have already been developed. Especially impressive is the city's newly prepared Biodiversity Strategy and draft Energy Plan. Sustainable energy policy (LSDC, 2003) which aims to increase resource efficiency, reduce anthropogenic climate change, tackle fuel poverty and boost new green industries is an important guideline for achieving a sustainable society. The draft energy strategy has created a solid foundation for a sustainable London by making these themes central to its approach. Biodiversity Strategy (LSDC, 2002) seeks to ensure that there is no overall loss of wildlife habitat in London, and that more open spaces are created and made accessible, so all Londoners are within walking distance of a quality natural space.

The Mayor of London established the London Sustainable Development Commission in 2002 with representation from economic, social, environmental, and London governance sectors, to present sustainability issues to the London government. The Commission's mandate is to provide a better quality of life for all Londoners.

In 2003, the Commission published *A Sustainable Development Framework for London* which gave decision and policy makers a list of fourteen objectives to achieve with any strategy, policy or project they wished to develop. These fourteen objectives relate to the four areas of sustainable development:

1) Taking responsibility for the influence of ones actions on other people and the environment, and thinking longer term objectives;
2) Developing respect for London's various communities and for London's environment;
3) Managing resources more cautiously to reduce London's environmental impact;
4) Getting results which achieve public, financial, social and environmental objectives and at the same time improve the quality of life for Londoners now and in the future.

To examine how London is achieving these objectives, the Commission has classified 20 headline Quality of Life indicators into the four key themes of Responsibility, Respect, Resources and Results. These four "R"s are London's principles for sustainable development. Responsibility includes consideration, ownership, information, participating, involving and long term thinking. Respect includes people, places, nature, diversity, vibrancy and safety. Resources list protection, recycling, environment, efficiency and prudence. Results incorporates transparency, innovation, improvement, accessible and well-being. (LSDC, 2005)

The above strategies provide a benchmark to measure whether government actions are making London a better place to live. London has the opportunity to become a cultured, balanced and sustainable city because people who live there developed a successful strategic body to fulfill its full potential. In the following statement, the parameters of the qualities of an eco-city system will be used to examine the effectiveness of these policies.

Planning for the 2012 Olympics, London has promised the world to make zero waste Olympic Games by maximizing recycling and minimizing waste materials and energy sources. The government will continue its role in sustainable development which is already being delivered by individuals and communities. London is moving towards becoming a truly sustainable city.

All in all, although cities have a choice of different approaches and strategies to become ecological, in terms of location, context, and city characteristics sustainability is still the long-term goal. This article draws two conclusions. The first conclusion is that each of these two cities needs to develop its own unique sustainable activities. Curitiba's, rubbish collection produces the highest recycling in the world - 70% - and the compact transportation system has reduced car traffic 30%. The city also has the largest downtown pedestrian shopping area in the world. These factors create a people-friendly city.

For London: the whole policy strategy is the right direction for the city to achieve successful sustainability; the city's transport network is the most efficient one in the world. The government encourages green design for buildings to minimize environmental damage. The second conclusion is that the dense city model and the qualities of eco-system can be used to evaluate most cities, even if they use different strategies to achieve sustainability.

REFERENCES

Archibugi, F. (1997). *The ecological city and the city effect: essays on the urban planning requirements for the sustainable city*, Ashgate Press.

Barton, H. (2000). *Sustainable communities: the potential for eco-neighbourhoods*, Earthscan Publication Ltd.

Burton, E., Jenks, M. and Williams K. (2000). *Achieving sustainable urban form*, E & FN Spon.

Girardet, H. (1992). *Creating sustainable cities*, Green Books Ltd.

Girardet, H. (1996). *The Gaia atlas of cities: New directions for sustainable urban living*, Gaia Books Limited.

Girardet, H.t (2004). *Cities people planet*, Wiley-Academy.

Gosling, D. and Maitland, B.(1984). *Concept of urban design*, London: Academy Edition.

Gumuchdjian, P. (1997). *Cities for a small planet*, Faber and Faber Limited.

Hall, Peter (1988). *Cities for Tomorrow: An intellectual history of urban planning and design in the twentieth century*, Basil Balckwell.

Hall, P. and Pfeiffer, U. (2000). *Urban future 21: A Global 21th Century Cities*, E&FN Spon.

Herbert T. D. and Thomas J. C. (1982). *Cities in space, city as place*, David Fulton Publishers Ltd.

SECTION V: CONCLUSIONS

In: Eco-City and Green Community
Editor: Zhenghong Tang

ISBN: 978-1-60876-811-0
© 2010 Nova Science Publishers, Inc.

Chapter 11

CONCLUSIONS

Zhenghong Tang

Community and Regional Planning Program
313 Architecture Hall, College of Architecture
University of Nebraska, Lincoln, NE, USA

This book provides theoretical discussions and historical overviews for eco-city and green community. Based on the above chapters and research papers, we have arrived at the following suggestions for achieving the goals of eco-city and green community.

First, this study provides a conceptualized definition for eco-city and green community which includes the following ten core aspects:

1) An eco-city and green community needs to protect and preserve the natural environment and undisturbed sensitive natural resources.
2) An eco-city and green community needs to support local agriculture and local products to reduce the ecological footprint at the community level.
3) An eco-city and green community needs to encourage a mixed-use, compact, clustered urban land use development pattern.
4) An eco-city and green community needs to develop a transit-oriented, pedestrian-oriented community by building walking and bicycling paths and promoting public transit use.
5) An eco-city and green community needs to develop efficient transport and communication systems to change traditional human behaviors. New technologies can reduce costs, improve efficiency, and save energy.
6) An eco-city and green community needs to maximize renewable energy and adopt energy conservation strategies.
7) An eco-city and green community needs to adopt zero-waste programs to recycle materials.
8) An eco-city and green community needs decision makers who understand the concept of ecosystem and sustainable development.

9) An eco-city and green community needs to promote stakeholder involvement to develop community-based efforts to address community quality.

10) An eco-city and green community needs to encourage public participation to support this campaign.

Based on the discussions in this book, we recognize that it is important to reform urban forms and structures. It is time to make dramatic changes in traditional urban planning in order to reach the goals of eco-city and green community. Minimizing the demand for land is the direct way to reduce urban sprawl. It is a myth that a city with more green fields means a greener and more ecologically friendly city; however, this idea may be wrong if considered in a regional context. More green fields generally lead to lower density which is a major factor for urban expansion. A mixed use, compact urban development form can minimize energy and material consumption and impairment of the natural environment and human health by creating a livable, safe, aesthetic, diverse, equal community.

It is also critical for an eco-city and green community to improve its transportation system by minimizing transport demand with an efficient public transit system. Urban planning must satisfy transit and mobility needs. Urban planning can improve accessibility to services and existing transport networks. An effective transportation system also reduces greenhouse gas emission, air pollution, and accidents.

A recent hot topic in eco-city and green community development is the creation of a more efficient energy and material flow. It is important to minimize material and energy consumption by using new green technologies. Energy system improvement should also improve health by improving indoor air quality and convenience of heating and cooling systems.

Comprehensive, dynamic, systematic thinking is the key to improving our understanding of eco-city and green community and changing the existing structure of urban planning and the performance of city planners. We believe that the campaign for eco-city and green community helps to build a balanced society with environment, social and economic sustainability. Eco-city and green community campaigns need to integrate bottom-up and top-down approaches. Our communities must make significant efforts in the following three areas: 1) collaborative support for initiatives, 2) bottom-up environmental stewardship and public behavior changes, and 3) new techniques, tools, and strategies.

In summary, an eco-city or green community should be environmentally friendly, socially equal and self-sufficient in energy, water and food production. Besides changes in the physical environment, it is important to consider the characteristics of production modes, consumption behavior and decision instruments. The campaign for eco-city and green community should have a comprehensive, systematic vision and promote collaboration with urban planning, transportation planning, public health, housing policy, energy policy and technologies, natural resources management, and social justice. In order to gain a holistic view of eco-city and green community, it is necessary to integrate science, technology, and policies. In particular, a multidisciplinary approach is needed for an effective collaboration to address this issue. For instance, asking what are the links between walkable community and climate change mitigation, and how to improve the understanding of greenhouse gas emission is of the outmost importance for urban planners and decision makers. The findings of this book will hopefully provide an opportunity and comprehensive perspective to find solutions for future sustainable communities. The concept of eco-city and green community presents the perfect image for our cities, but the road leading to that goal is difficult. The strategies for building eco-city and green community are different because cities and communities vary in size, scope, needs, capacities, resources, and commitment. There is no standard label for eco-cities and green communities.

Although this book has advanced the understanding of the theories and practices in eco-city and green community, it has several limitations. While this book has provided a theoretical review of planning theories, it is a primer for research to investigate the topic. It was a challenge to decide which theories to explore in regard to eco-city and green community since many theories actually interact; therefore, it is difficult to conclude which theory is better than the others. Although this book has conceptualized some characteristics and key elements for eco-city and green community planning, an actual protocol which can be directly used in the planning field is still distant. We also recognize that the practice of eco-city and green community is a dynamic, developing process. This book has discussed a series of approaches, strategies, tools, and policies for eco-city and green community development which are based on individual cases rather than comprehensive discussions. Furthermore, if possible, this book would have involved more cases (Waitakere in New Zealand) to analyze the effectiveness of these theories overseas. The discussion and comparison of the urban form models may be subject to a series of factors such as geographical variations, socioeconomic characteristics, and policy framework. This book is the starting point for discussing major theories and practical cases. Future research and empirical studies will be necessary to analyze planning theories in different cities. These studies can then be summarized and added to the established planning theories.

CONTRIBUTORS

Editor, **Zhenghong Tang** is an assistant professor in the Community and Regional Planning Program in the College of Architecture at University of Nebraska-Lincoln. He is a courtesy assistant professor in the School of Natural Resources, a faculty fellow in the Water Center and the Center for Advanced Land Management Information Technologies at University of Nebraska-Lincoln. He is also a research fellow in the Environmental Planning and Sustainability Research Unit in the Hazard Reduction and Recovery Center at Texas A&M University. His previous research addressed watershed management information system, environmental spatial modeling, soil erosion and non-point pollution mechanisms, geohazard risk assessment, and land use driven-factors analysis. His recent research mainly emphasized integrating strategic environmental impacts with local land use planning. His current research and outreach projects focus on local land use planning capacity, natural resources management, hazards mitigation, and climate change mitigation and adaptations. He teaches undergraduate and graduate courses in Environmental Planning and Policy, GIS in Environmental Design, and Planning Theory. His research and teaching interests cover: land use planning, environmental planning & policies, environmental impact assessment, local climate change responses, urban and regional development policies, environmental hazards management, watershed management, quantitative methods and GIS analysis, and international (China) planning. (ztang2@unl.edu)

CONTRIBUTING AUTHORS

Ting Wei, University of Nebraska-Lincoln (weitingkoo@gmail.com)
Jeffrey R. Kenworthy, Curtin University, Australia (j.kenworthy@curtin.edu.au)
Szu-Li Sun, University of Reading Business School, UK (klr02ss@yahoo.com)
Yunwoo Nam, University of Nebraska-Lincoln (ynam2@unlnotes.unl.edu)
Erin Bolton, University of Nebraska-Lincoln (erin.a.bolton@gmail.com)
Shaojing Tian, University of Nebraska-Lincoln (tianshaojing@yahoo.cn)
Akhilesh Surjan, United Nations University, Japan (akhilesh.surjan@yahoo.com)
Prasanta K. Mohapatra, United Nations University (prasant_mohapatra @hotmail.com)
Praveen K. Maghelal, University of North Texas (Praveen.Maghelal@unt.edu)

INDEX

D

E

Q

T

U